# BEYOND NATO

# BEYOND NATO

## Staying Out of Europe's Wars

# TED GALEN CARPENTER

CATO
INSTITUTE
Washington, D.C.

**Library of Congress Cataloging-in-Publication Data**

Carpenter, Ted Galen.
  Beyond NATO : staying out of Europe's wars / Ted Galen
Carpenter.
    p.     cm.
  Includes bibliographical references and index.
  ISBN 1-882577-16-7. — ISBN 1-882577-17-5 (pbk.)
  1. North Atlantic Treaty Organization.   2. Europe—Defenses.
3. National security—United States.   I. Title
UA646.3.C269   1994
355'.031091821—dc20                                           94-23704
                                                                    CIP

Cover Design by Colin Moore.

Printed in the United States of America.

CATO INSTITUTE
1000 Massachusetts Ave., N.W.
Washington, D.C. 20001

To my mother, Magdalene Carpenter, who instilled in me a love of learning and a determination to always question the conventional wisdom.

# Contents

# Acknowledgments

I owe a debt of gratitude to many people who helped make this book possible. Ed Crane, president of the Cato Institute, has consistently provided moral and tangible support for my efforts to challenge the foreign policy status quo. He has never wavered in his understanding that the values of limited government and individual liberty cannot flourish at home if the United States continues to pursue an expensive and dangerous interventionist policy abroad.

Jonathan G. Clarke, Ronald Steel, Alan Tonelson, Eric Nordlinger, Benjamin C. Schwarz, and Barbara Conry provided useful comments and suggestions on an earlier draft of the book. Meelis Kitsing spent many hours assisting with the research, including tracking down several elusive sources. David Lampo worked diligently to keep the project on schedule. My editor, Elizabeth W. Kaplan, deserves special credit for polishing the style and otherwise improving the quality of the manuscript.

Most of all, I want to thank my wife, Barbara, for her support and patience, even when the creative process consumed large portions of my time and attention.

# Introduction

When Stalin's curtain came down with a clank,
Stern NATO guarded Western Europe's flank.
The cold war's gone, but no one seems to think
That NATO should disband, or even shrink.
The problem NATO's facing—and it's vast—
Is how far to expand, and just how fast.
The Poles? The Czechs? Arrangements can be made
For them to join before the next parade.
The other former Reds could not abstain.
And soon we'd have, with all its bombs, Ukraine.
Which means in time our favorite boychik, Boris
Would join, and bring the Russian Army Chorus.
United, NATO troops could stand foursquare
Against an armed aggression launched from . . .
     where?

—Calvin Trillin,
"NATO in the New World Order"

For a brief time there was speculation in the United States and Western Europe that the collapse of the Soviet Union's East European empire, followed by the disintegration of the USSR itself, might signal NATO's ultimate demise as well. It was difficult to envision an alliance that had been created to thwart the expansionist ambitions of a totalitarian superpower playing a meaningful role in the absence of such a threat. NATO, it seemed logical to assume, would follow its Cold War organizational adversary, the Soviet-dominated Warsaw Pact, into oblivion.

---

But NATO's defenders have worked hard to build the mystique that the alliance is even more important in the post–Cold War era than it was during the Cold War. Not only do they insist that NATO is essential to prevent the resurgence of the instability and national rivalries that spawned two world wars in Europe, they argue that Washington's leadership role in the alliance, symbolized by the continued deployment of American troops on the Continent, is imperative as well. NATO is the only institutional vehicle for U.S. influence in European affairs, Atlanticists contend; any U.S. retrenchment on security issues would, therefore, jeopardize important American political and economic interests.

Instead of considering whether an alliance created to wage the Cold War is relevant in the vastly altered setting of post–Cold War Europe, policymakers are debating whether NATO should enlarge its membership by incorporating some or all of the Central and East European nations. Proposals to expand NATO's security role eastward have become a growth industry on both sides of the Atlantic. Central and East European political leaders are lobbying vigorously to have their countries admitted to the alliance, and they have enjoyed considerable support from the German government; the late Manfred Wörner, NATO's secretary general; and other influential figures.

Several prominent U.S. politicians, pundits, and foreign policy experts have published articles during the past two years advocating the expansion of NATO's membership, or at least its security jurisdiction.[1] The underlying rationale for such policy prescriptions is that NATO's post–Cold War mission should be to promote greater stability throughout the Continent, dampen the conflicts that are now bedeviling portions of Eastern Europe, nurture the fragile democracies that have emerged from the wreckage of the Soviet empire, and prevent the emergence of an expansionist threat similar to those once posed by Nazi Germany and the Soviet Union. Advocates of enlargement secured an important endorsement in early 1994 when the U.S. Senate voted 94 to 3 to approve a resolution favoring the expansion of the alliance.[2] Although that resolution was nonbinding, it had major symbolic value: the margin of passage demonstrated that the goal of an expanded NATO has widespread support in both political parties.

The principal obstacle to expansion has been the concern that Moscow might regard the move as a hostile act, which could damage the West's relations with the newly democratic Russia. Proponents of

2

enlargement were buoyed in August 1993 when Russian president Boris Yeltsin indicated that his government would not object if Poland or the Czech Republic sought to join the alliance. The Russian government's rapid backtracking on the issue, however, dampened the enthusiasm for enlargement and confirmed the views of critics who warned that Yeltsin's accommodating attitude was not shared by Russia's political and military hierarchy.

Both Yeltsin and Foreign Minister Andrei Kozyrev have subsequently emphasized that the only version of NATO's expansion that would be tolerable to Moscow is one that included Russia. Although some Western advocates of enlargement consider that course of action acceptable, others warn that such broad expansion would dilute NATO's security role to the point of impotence. The East Europeans are even more insistent that the main purpose of a broader NATO must be to guard against the possibility of a new round of Russian expansionism; in their view, a NATO that included Russia would automatically jeopardize that mission.

The Clinton administration has thus far declined to endorse the most ambitious expansion schemes, fearing that a premature enlargement of NATO could antagonize Russia and perhaps even undermine the Yeltsin government by giving Vladimir Zhirinovsky and other ultra-nationalists an appealing issue to exploit. (Indeed, the administration has come under fire from some NATO enthusiasts for having a Russophilic foreign policy because of the failure to press forward on the enlargement issue.)[3] As an alternative to a program for immediate expansion, administration officials offered a vaguely defined security partnership—the Partnership for Peace—between NATO and its eastern neighbors.

In the long run, that may be a distinction without a difference. Secretary of Defense Les Aspin, the principal author of the partnership concept, stated openly that it should be a first step toward possible alliance membership for the old Soviet bloc states, and other administration policymakers have made similar comments.[4] President Clinton himself embraced the long-term goal of a larger NATO during a visit to Prague in January 1994. "While the partnership is not NATO membership, neither is it a permanent holding room," Clinton emphasized. "It changes the entire NATO dialogue, so that now the question is no longer whether NATO will take on new members, but when and how."[5]

Although some current NATO members (most notably Britain and Italy) are wary of expanding the alliance, given the volatile political environment throughout Eastern Europe, the pressure is building to enlarge NATO's security jurisdiction in some fashion.[6] Wörner echoed the views of U.S. officials in stressing that the Partnership for Peace should not be viewed as an alternative to eventual NATO membership for the former Soviet bloc states, adding that he sensed "a growing consensus that, in principle, the alliance should open up to new members."[7] Although the alliance leaders followed the wishes of the Clinton administration and endorsed the Partnership for Peace at the NATO summit in January 1994, instead of offering full membership to the eager Central and East European candidates, that action appears to be merely the first stage of the expansion process. One immediate activity of the Partnership for Peace will be joint training and field exercises to strengthen military ties with the nations of Central and Eastern Europe.

Enlarging NATO's security responsibilities is a dubious step for numerous reasons. If Russia is excluded from membership, expansion schemes have a significant potential to provoke that country. Even democratically inclined Russians would have reason to be suspicious of a powerful, U.S.-led alliance actively involved in a region in which Russia has long-standing political, cultural, economic, and strategic interests. Yet if Russia were included in the expanded roster of members, NATO would become an unwieldy entity spanning every time zone on the planet.

Perhaps even more troubling, expansion would convert an alliance designed to achieve clear and limited security objectives in a relatively stable Cold War setting into a nebulous crisis-management organization in a highly unstable post–Cold War setting. NATO would change from a defensive alliance to protect the territory of member states from attack into an alliance to project force—a different mission with a vastly different set of risks and obligations. Even if the formal membership of the alliance were not increased, the de facto extension of NATO's security jurisdiction eastward would be fraught with danger. NATO would become entangled in an assortment of disputes among the Central and East European states, many of which have the potential to explode into armed conflict at any time. Given the fragile nature of some of those nations, the alliance could even become mired in a number of internecine struggles.

That danger is not merely hypothetical. NATO is already involved militarily in the Bosnian civil war, and that involvement appears to be escalating. Not only do several NATO members have troops in that country as part of the UN peacekeeping force, but NATO ships in the Adriatic help enforce the economic embargo against Serbia and NATO aircraft enforce the no-fly zone over Bosnia itself. In early 1994 the alliance increased the scope of its intervention by establishing exclusion zones around Sarajevo and other Muslim-held cities in Bosnia and demanding that Bosnian Serb forces cease their attacks on such safe havens. NATO's increasingly punitive policy toward the Serbs has, in turn, exacerbated tensions with Russia, which has regarded the Serbs as political and religious allies for generations.

There are indications that the alliance's involvement may deepen regardless of the flow of events in Bosnia. In mid-July 1994 U.S. secretary of defense William Perry asserted that the West was "at a fork in the road. I think that if we go down either one of those two forks . . . it would be an expanded role for NATO, and the United States would be an important part of that."[8] Perry explained that if the parties to the Bosnian war reached a peace settlement, the NATO powers would contribute additional troops to the UN peacekeeping force to enforce that accord. Conversely, if the fighting—which had been relatively subdued in the late spring and early summer—flared again, NATO would probably expand military operations to protect the Muslim safe havens established by the UN Security Council.

The Bosnian operation is a model of the missions NATO will be obligated to undertake if it moves eastward. If advocates of expansion are serious about the role NATO should play in helping to stabilize and pacify Eastern Europe, Bosnia will simply be the first of many such missions. The various difficulties the alliance has encountered in Bosnia are also likely to be replicated elsewhere. The murky nature of the conflict itself, the dim prospects for a definitive settlement, the use of NATO's military resources in the pursuit of nebulous political objectives, the prospect of a temporally open-ended commitment, and the policy disagreements and exacerbation of tensions between NATO and Russia are all indicative of what the alliance can expect in other peacekeeping enterprises throughout Eastern Europe. That prospect should be of more than academic interest to Americans, because the United States, as NATO's acknowledged leader, is certain to be deeply involved in such activities.

During the Cold War, NATO's primary purpose was straightfor-
ward: to deter Soviet aggression against Western Europe. To be sure,
secondary objectives existed, such as constraining German power
within a multilateral institution and fostering Western political and
economic cooperation, but thwarting Soviet hegemony was the alliance's raison d'être. Most Americans at least tacitly accepted the risks
entailed by Washington's alliance obligations, fearing that otherwise
the Soviet Union—which already posed a serious threat to America's
security—would gain control of the population and technology of
Western Europe and become an even greater menace.

NATO's proposed new missions—to stymie the reemergence of a
Russian sphere of influence in Eastern Europe and to prevent general
"instability" in the region—do not have comparable relevance to
America's security interests. With the demise of the Soviet Union, the
United States faces no credible great-power threat to its well-being, nor
is it likely to in the foreseeable future. Acceptance of a Russian sphere
of influence would be little more than recognition of one of the most
enduring realities of international politics: that great powers typically
create geopolitical buffer zones to protect their own security and
advance their economic and political agendas. The establishment of
such a zone does not necessarily portend the kind of aggressive
expansionism that could threaten the global configuration of forces
favorable to American interests. Moreover, given its relatively modest
military strength and its daunting political and economic problems,
Russia is not likely to make a bid for global hegemony for many years
to come, if ever. A Russia shorn of its empire and not ruled by an
aggressive totalitarian regime would not inherently threaten America's
security merely by acting as a normal great power and carving out a
limited sphere of influence in Eastern Europe.

Undertaking an obligation to help keep Eastern Europe out of
Moscow's orbit would constitute a radical departure in U.S. foreign
policy. Despite the intense Cold War rivalry with a superpower
challenger, no American administration, liberal or conservative, Democratic or Republican, was willing to risk war to roll back Soviet power
in the region. Yet those who would have NATO become the security
shield for Eastern Europe are apparently willing to incur the risk of a
clash with a nuclear-armed Russia to prevent the establishment of a far

less menacing zone of Russian influence in that region. How the assumption of such a risk would be in the best interest of the United States is not readily apparent.

The proposed stability mission for NATO is even less germane to legitimate American security interests. There is already ample evidence that Eastern Europe may well be a turbulent and disorderly region in the post–Cold War era, but that is an argument for American caution and aloofness rather than intervention. Instability per se does not threaten the United States. Conflict and disorder have always been present in the international system, but only a small percentage of such disputes have the potential to menace America. The conflicts in the Balkans and along the perimeter of the former Soviet empire may be important to the parties involved—and perhaps to their immediate neighbors—but they are nevertheless largely parochial. Most of them do not have the potential to disrupt even the European, much less the global, balance of power. Extending NATO's security jurisdiction would entangle the United States in the myriad ethnic, religious, and territorial disputes of Central and Eastern Europe without any compelling need or prospective benefit.

Sen. Richard Lugar (R-Ind.) summarized the view of many supporters of NATO when he stated candidly that the alliance must "go out of area" or "it will go out of business."[9] But if NATO has accomplished the mission for which it was designed, then it is reasonable to ask whether it should go out of business. The responsibility of U.S. policymakers is to protect the security interests of the American people, not to preserve an alliance for its own sake. However much entrenched political, economic, and intellectual vested interests on both sides of the Atlantic may want to perpetuate a Cold War institution that benefits them, a policy of "NATO forever" is not necessarily wise or even sustainable. If NATO can continue to exist only by adopting the dangerous and vaguely defined mission of crisis management in Eastern Europe, Americans have every right to wonder whether it has outlived its usefulness.

Today, NATO is an alliance in search of a purpose. Indeed, the proliferation of expansion schemes appears to have as much to do with the politics of institutional self-preservation as with the defense of bona fide security interests.

Beyond the expansion issue, which has dominated the debate about the future of NATO, lies a far more important question that needs to be addressed. The premise that originally underlay the alliance was the fundamental compatibility of interests between the United States and its European allies. Although that premise seemed convincing during the Cold War, when the Western democracies faced a powerful common adversary, the validity of assumptions about transatlantic solidarity is far more questionable in the post–Cold War era. Indeed, there is growing evidence that the economic, political, and even security interests of the United States and the major powers of Western Europe are diverging in ways both obvious and subtle. The pertinent question for U.S. policymakers is whether it makes sense from the standpoint of American interests to preserve a transatlantic alliance that was designed in a vastly different era to deal with a mutual threat that no longer exists. That point is important, not only because America's continued leadership of NATO involves significant financial costs—some $90 billion a year—but because the new missions being contemplated for the alliance would entail grave risks.

American leaders should not only resist suggestions to enlarge NATO's security jurisdiction, they should seriously consider a policy that moves in precisely the opposite direction—toward giving the venerable alliance a well-earned retirement. As part of that reassessment, Washington must recognize that the concept of an Atlantic community, whatever validity it may have had during the Cold War, no longer has the same relevance. Although European and American security interests may overlap, they are, nevertheless, distinct and sometimes may even be in conflict. Rather than strive to preserve an outdated transatlantic alliance, Washington would be wise to encourage the European states to form new security structures or to strengthen such existing bodies as the Western European Union and the Franco-German Eurocorps as replacements for NATO.

Such "Europeans-only" organizations would bring to an end Western Europe's military reliance on the United States. Although that would probably mean a dilution of American influence in European affairs, the benefits would outweigh the drawbacks. Devolving responsibility for their own security to the West Europeans would resolve the problem of their "free riding" on the U.S. security guarantee—an annoying phenomenon that has cost American taxpayers hundreds of billions of dollars during the history of the alliance and sparked numerous acrimonious disputes about "burden sharing." Phasing out NATO would also end Western Europe's unhealthy dependent mind-set on security issues, which has rendered the European powers so tentative and ineffectual in dealing with the region's disputes in the post–Cold War period. The major European powers would have to take responsibility for security problems instead of always looking to the United States to provide solutions. Finally, Europeans-only organizations would be more appropriate for dealing with the purely local or regional conflicts and quarrels of post–Cold War Europe.

An important initial step is to change the nature of the debate about NATO's future. Too often the debate has been a narrow and intellectually sterile discussion of when, how, and how far the alliance should expand. The more fundamental issue is whether the alliance should exist at all. A high-priority item on Washington's foreign policy agenda should be to move beyond a reflexive reverence for NATO toward a new European policy that better serves America's interests in the post–Cold War era.

# 1. The Campaign for an Enlarged NATO

The principal impetus for expanding NATO has come from two sources: the political leaders of virtually all Central and East European countries and prominent members of the U.S. foreign policy community, including some high-profile politicians. Conversely, a majority of the political leadership in Western Europe has thus far been rather cool toward the idea of a larger NATO, and even the foreign policy communities in those nations appear to be ambivalent. Although the U.S.–East European axis has been able to mount an impressive lobbying effort, the objectives of the two factions differ in crucial ways that in the long term are likely to cause problems for the campaign to expand the alliance.

The goals of the East Europeans are unambiguous; they want early membership in NATO and the full protection of the alliance's security guarantees. Czech president Vaclav Havel expressed the view of many of his colleagues when he urged the alliance to consider his country, Poland, Hungary, and Slovakia (the so-called Visegrad powers) "prime candidates for early membership," with the other "post-Communist countries" brought in at a later date.[1] Although the East European states continue to hope for an affiliation more substantial than the Clinton administration's Partnership for Peace, they embrace even that limited measure. The Visegrad powers, the Baltic republics, and other East European states have signed the presentation documents for the Partnership for Peace and have agreed to participate in joint military training exercises. Even a marginal candidate for membership, Romania, stresses that it is configuring its military to be more compatible with NATO, looks forward to joint exercises, and anxiously awaits the day when it can become part of NATO's military arrangements.[2]

In most of their public statements, the East European leaders avoid identifying potential enemies, preferring instead to cite general security concerns. Thus, Laslo Kovacs, chairman of the Hungarian parliament's Foreign Affairs Committee, states, "The security risk we now face stems from the instability of the region rather than a traditional military threat."[3]

Despite such rhetoric, East European officials have rather specific security threats in mind. Privately, some leaders still express fear of Germany's long-term ambitions, although even the most paranoid concede that the notion of a Fourth Reich is rather far-fetched at the moment. A much more pressing concern to the East Europeans is Russia. Havel offers a revealing hint of the principal source of his worry when he discusses how the West should react to the emergence of a new hard-line government in Moscow. If such a regime came to power, the West would "make the most fatal possible mistake" if it adopted a policy of appeasement, Havel warns. The East Europeans' fear of a new wave of Russian expansionism was evident even before the surprisingly strong showing of Vladimir Zhirinovsky's ultranationalist Liberal Democratic party in Russia's December 1993 parliamentary elections. Since then, their desire for NATO protection from a neo-imperial Russia has intensified.

Poland's president Lech Walesa has been the most outspoken in warning of Russian revanchism and in seeking the shelter of NATO's security umbrella. During President Clinton's visit to Warsaw in July 1994, Walesa stated that he was 60 percent apprehensive about peace with Russia and only 40 percent optimistic that Moscow will not order troops into Poland at some point. An American military, as well as an economic, presence was necessary, he contended, to change that pessimistic assessment.[4]

Whereas the East European countries have quite specific security objectives, U.S. proponents of an enlarged NATO often speak in the most ethereal terms about fostering political cooperation and promoting stability. Only a few are candid enough to state publicly that NATO's expansion eastward would be directed against Russia and designed to prevent the rebirth of the Soviet empire. Some of that vagueness may be a deliberate tactic by foreign policy activists. Even the most enthusiastic partisans of a larger NATO are wary of admitting that expansion would be for the purpose of waging a second Cold War. Such an prospect might alarm the American public, which of late has been noticeably unenthusiastic about embracing dangerous new overseas commitments—as the reaction to U.S. military casualties in Somalia and the continuing resistance to proposals to intervene militarily in Bosnia demonstrate.

But there appear to be more profound reasons for the gossamer nature of the expansion proposals coming from American supporters of NATO. Unlike their East European brethren, who see NATO first and foremost as a military alliance, the Americans increasingly portray it as primarily a political association. That rationalization enables them to skirt discussions of the tangible costs and risks of expansion. At the same time, however, it causes the enlargement proposals to be not only vague but also shot through with internal contradictions. The "NATO as a political rather than a military organization" thesis also puts the American supporters of expansion on a collision course with their East European friends.

U.S. partisans of an enlarged NATO invariably contend that the alliance has a vital role to play in preserving the Continent's security and political stability in the post–Cold War era and that Eastern Europe will be the most important region for the pursuit of that mission. Former secretary of state James Baker III stated bluntly, "The alliance's post-Soviet imperative is clear: to extend democratic values to and protect Western interests in the East."[5] RAND Corporation analysts Ronald D. Asmus, Richard L. Kugler, and F. Stephen Larrabee agree.

> Nationalism and ethnic conflict have already led to two world wars in Europe. Whether Europe unravels for a third time this century depends on if the West summons the political will and strategic vision to address the causes of potential instability and conflict before it is too late. A new U.S.-European strategic bargain is needed, one that extends NATO's collective defense and security arrangements to those areas where the seeds for future conflict in Europe lie: the Atlantic alliance's eastern and southern borders.[6]

Jeffrey Simon, senior fellow at the National Defense University's Institute for National Strategic Studies, stresses similar themes. If the new democracies of Eastern Europe "are not soon given a hope of eventual security within a broader NATO," Simon warns, "they may come to feel rejected, to look elsewhere, or to succumb to internal reactionary forces." Efforts to create liberal democracies in the region may then fail, "and much of the West's investment in the cold war will be squandered."[7] Simon summarily rejects the notion that, with the end of the Cold War, NATO is obsolete. "NATO is not an organization whose mission is over. It continues to perform vital European security roles in establishing regional stability, embedding Germany in a multilateral transatlantic security structure, preventing re-nationalization in European force planning, anchoring the flanks of Europe, and providing a credible capability for potential out-of-area actions."[8] To perform such roles effectively, he insists, the alliance must expand eastward.

One striking characteristic of the various American proposals to enlarge NATO is their haziness on several crucial issues. The advocates are vague, if not evasive, about which nations should be admitted as members, the pace of expansion, and the scope of the security guarantees that should be given to new members. In light of the impact that a dramatic transformation of NATO's composition and mission would have on the U.S. role in Europe, and potentially on the nation's security and well-being, such fuzziness is cause for concern.

Proponents of expansion generally agree that integration into NATO should be offered first to Poland, Hungary, and the Czech Republic—three of the four Visegrad countries. That is where the clarity ends. University of Chicago professor Charles L. Glaser, for example, contends that the West has reasonable grounds for extending security guarantees to those nations (as well as Slovakia) but not making such commitments to the states of the former Soviet Union.[9] Hudson Institute research fellow Gary L. Geipel also would enlarge NATO to include "three or four" Central European countries but is unenthusiastic about incorporating their eastern neighbors.

> [NATO] cannot expand much beyond a certain size without sacrificing so much efficiency and readiness that it becomes dysfunctional. A NATO of nineteen or twenty members, as opposed to the current sixteen, might already approach or exceed the limits of its viability. If NATO is to work for anyone, it cannot work for everyone. That consideration will keep it from expanding much further.[10]

In contrast, Simon would ultimately open the ranks to all members of the North Atlantic Cooperation Council—the consultative body established by NATO in November 1991 to serve as liaison with the nations of the former Warsaw Pact—after a probationary "associate" membership.[11] That could mean that the Baltic states, Ukraine, and even Russia would eventually be eligible to join NATO. Asmus, Kugler and Larrabee stress the desirability of incorporating the Visegrad states but are utterly vague about whether similar offers should be made to some or all of the other Central and East European states. Edward L.

Rowny, who served as special adviser to Presidents Reagan and Bush on arms control issues, likewise suggests offering full membership in the alliance to Poland, Hungary, and the Czech Republic but adds vaguely that "Eastern European nations—Russia, Ukraine, Belarus, and the Baltics—should be given additional support, outside of NATO channels," to protect their security and encourage further political and economic reform.[12]

NATO's political boosters in the United States are equally vague and divided on the scope of expansion. Sen. Richard Lugar (R-Ind.) would bring in Poland, Hungary, and the Czech Republic promptly, and he apparently hopes to make membership available at a later date to the other states of the region. But he waffles when pressed to indicate how far east NATO should go, touting the virtues of a "flexible" approach. The status of Ukraine, he admits, is an especially difficult problem. The West can either support a neutral Ukraine based on the Finnish or Swedish model, or "the West can bite the bullet and discuss real and meaningful security guarantees" for that country.[13] James Baker, Jeane Kirkpatrick, and other political leaders suggest comprehensive expansion, even offering eventual membership in the alliance to Russia and the other former Soviet republics. During a visit to Kazakhstan in October 1993, Secretary of State Warren Christopher spoke casually about that country's participation in NATO military exercises as a step toward eventual membership.[14] Kazakhstan and other Central Asian republics, such as Turkmenistan, have already become members of the Partnership for Peace, and they express an interest in becoming full-fledged members of the alliance.

## Diverse Expansionist Factions

Three reasonably distinct factions in the campaign to expand NATO have emerged from all the intellectual ferment—although the placement of specific individuals is not always easy, given their hazy public writings and comments. The first group consists of the anti-Russian limited expansionists. Its adherents fear a revival of Moscow's imperial ambitions and want to improve the conventional military position of the West by moving NATO's frontiers eastward. At the same time, they implicitly or explicitly acknowledge that not all Central and East European nations are ready for NATO membership and that some

15

would not add materially to the alliance's strength. They also concede that some additions would have a greater potential than others to provoke a rash Russian response.

Although the limited expansionists would incorporate the Visegrad powers, they exhibit little enthusiasm about further expansion. (Indeed, some even omit Slovakia from their list of acceptable candidates—primarily because of that country's dubious commitment to democracy and its continuing economic stagnation.) Former secretary of state Henry Kissinger; Lt. Gen. William E. Odom, former director of the National Security Agency; and Edward Rowny are among the notables who appear to be members of the anti-Russian limited expansion faction—although even they usually make pro forma statements that expansion would not be directed against Russia or any other power.

The second faction consists of the anti-Russian broad expansionists.  Its members share the view of the limited expansionists that renewed Russian imperialism is a real danger and that the West should take advantage of Moscow's current weakness to improve NATO's strategic position. Where they part company with their more cautious brethren is in the scope of enlargement. Believing that a second Cold War is virtually inevitable, they propose taking in not only all of the Soviet Union's one-time Warsaw Pact satellites but at least some of the former Soviet republics, including the Baltic states and Ukraine, as well. Such expansion, they assume, would create an impregnable western barrier to any effort by Russia to restore the empire. The most outspoken proponent of the anti-Russian broad expansion position is former deputy assistant secretary of defense Frank J. Gaffney, Jr., although the comments of former secretary of state Alexander Haig indicate that he might also fit into that camp. Under certain conditions—especially if tangible evidence emerges of neo-Russian imperialism—the RAND threesome of Asmus, Kugler, and Larrabee might also adopt that position.

The third faction consists of the all-inclusive expansionists. Although they may fear a revival of Russian aggression, they appear to be even more concerned about again dividing Europe militarily—with the principal differences being that this time the West would be responsible for that tragedy and that the new Iron Curtain would be a few hundred miles farther east. NATO enlargement motivated by anti-Russian animus, they also warn, could create a self-fulfilling prophecy as an alarmed and suspicious Russia reacts to what it perceives as hostile Western moves that threaten its legitimate security interests. To overcome such problems, advocates of all-inclusive enlargement would ultimately offer NATO membership to Russia as well as the East European nations and other former Soviet republics. The most prominent proponents appear to be former secretary of state James Baker, former UN ambassador Jeane Kirkpatrick, and such scholars as Simon.[15]

Advocates of all-inclusive expansion differ from their competitors in one very important way. Despite occasional boilerplate to the contrary, the other two factions clearly believe that NATO's principal relevance in post–Cold War Europe is as a military, not a political, association. Their motive in wanting to expand the alliance is to more effectively deter Russian aggression or, failing that, to improve NATO's ability to win the resulting military struggle. The all-inclusive expansionists, in contrast, see NATO primarily as a political organization for defusing or resolving conflicts throughout the Continent before they escalate into armed conflict. To the extent that they contemplate any military role for NATO, it would be as a collective security body—essentially a mini–United Nations—rather than as a classic military alliance directed against a specific nation or nations. Both aspects of their agenda—the scope of expansion and the purpose of the alliance—run directly counter to the goals of the East European governments who are the most eager to become members.

### Conflicting Agendas

The imprecision and lack of unity that proponents of expansion exhibit on the question of which nations should be invited to join NATO are equally evident on other important matters. One of the thorniest issues is the question of security guarantees. Article 5 of the North Atlantic Treaty declares that an attack on one member shall be considered an attack on all, and it obligates every signatory to render assistance to the treaty partner under assault. When NATO was first formed in 1949, Secretary of State Dean Acheson bluntly described article 5 as "the guts" of the treaty.[16]

Even many defenders of NATO are uneasy about obligating the existing members to defend the nations of Central and Eastern Europe. Yet admitting those states to membership would seem to entail such a commitment under article 5. In addition to questioning the wisdom of expanding article 5 obligations, some skeptics also wonder whether the countries of Central and Eastern Europe are ready for NATO membership. After all, their democratic systems are still in their infancy, there is massive economic instability, and it is not clear that their militaries would be able to make a meaningful contribution to the common defense.

Faced with such objections, several advocates of enlargement have tried to finesse the problem by suggesting an expansion that would—at least initially—be devoid of article 5 obligations. Simon, for example, promotes the concept of "associate membership" as a way station for new NATO members. The Central and East European countries would have the right to participate in many NATO deliberations and activities, but on security matters they would only have the right "to bring issues to the alliance under Article 4 of the NATO treaty."[17] Article 4 states that the parties to the treaty "will consult together whenever, in the opinion of any of them, the territorial integrity, political independence, or security of any of the Parties is threatened." (The Partnership for Peace, in fact, confers the right to such consultations to all nations that have signed the presentation agreements.)

Asmus, Kugler, and Larrabee adopt a similar approach to avoid the thorny issue of article 5 commitments, stressing that NATO membership "can come in phases and should be made conditional." That way NATO can supposedly "help solidify a zone of stability in Central Europe without undue risk of embroiling NATO's existing members in new ethnic or intra-regional conflicts." Association agreements would spell out the criteria for membership "but not provide explicit security guarantees."[18]

Such schemes may deserve high marks for creativity, but they are little more than exercises in wishful thinking. Politically, it is difficult to imagine how one could long sustain an alliance with two classes of members: a first class entitled to military assistance from its allies under article 5 and a second class that would receive only the thin gruel of article 4 "consultations" in the event of trouble. On that issue, there is a glaring conflict between the agendas of the Western proponents of expansion and the governments of the Central and East European nations that seek to join NATO. The former seem to regard enlargement primarily as a political exercise to somehow enhance European "stability." The latter, however, regard NATO as a lifeline to guarantee their independence from powerful adversaries—especially a revanchist Russia. To those governments and populations, NATO's great appeal is precisely the obligations of mutual assistance set out in article 5. Apprehensive Central and East European nations want reliable protection, not merely consultations.

The cool reception that some East European officials initially gave to Clinton's Partnership for Peace confirms that assessment. Hungarian defense minister Lajos Fur's reaction to the partnership concept was quite candid. "Hopefully it's a step toward NATO integration, which is our aim. But unfortunately it does not give us what we need—a guarantee of security."[19] During Clinton's visit to Prague in January 1994, after the NATO summit, Havel made that point more explicitly, stating that the Visegrad powers welcomed the Partnership for Peace as "a good point of departure." He added, however, that those countries "do not regard the Partnership for Peace as a substitute for [full NATO] membership, but rather as a first step toward membership."[20] Slovakia's government adopted a similar position in its memorandum formally accepting the Partnership for Peace.

> The Slovak Republic fully comprehends the gradual process of creating security mechanisms which ensure stability throughout the European continent, that has found its expression in proposals and programmes raised by the NATO summit. At the same time, the Slovak Republic deems it necessary to stress that it will proceed in its initiatives aimed to achieve full membership in NATO and the security guarantees which are a prerequisite for the successful accomplishment of its economic transformation and strengthening of its democracy.[21]

The inadequacy of consultations as an alliance benefit underscores the inherent problem with a bifurcated membership. Once NATO's boundaries were expanded, it would be exceedingly difficult to watch even an associate member be attacked without launching a collective military response. The very act of offering associate membership implies that the nation in question is important to the security of the existing alliance signatories—otherwise, the offer would never have been made. There would, therefore, be tremendous pressures to render military aid, whether or not an explicit article 5 obligation was in effect. One cannot have an alliance in which some members receive protection while others do not.

Associate membership may in fact be little more than a political pacifier to soothe American and West European publics into believing that NATO can be expanded without their countries' incurring new

19

risks. Even proponents of enlargement tacitly concede that a two-tiered alliance is not sustainable in the long run. Simon, for example, proposes that after a 5- to 10-year period of associate membership, a state would become eligible to apply for full membership. If the other NATO states approved that application, the country would then enjoy the guarantees of article 5. What he and other NATO expansionists decline to acknowledge is that the new treaty signatories would enjoy de facto article 5 protection from the outset.

It is probable that the East European nations will not be satisfied even with the paper guarantee of protection that comes with NATO membership, even if it includes article 5. Sooner or later—probably sooner—they will want NATO, including U.S., tripwire forces in their countries to make the treaty provisions credible. A subtle but revealing hint of that intent came from Hungary's ambassador to the United States during an interview with Free Congress Foundation president Paul Weyrich on National Empowerment Television. When asked whether his country would want Western troops as well as alliance membership as a tangible expression of NATO's security commitment, the ambassador replied merely that it was "too early" to consider that issue.[22] He could just as easily have stated that such a step would not be necessary, or that Hungary had no desire for any foreign troops on its soil, but instead he chose a response that had very different implications.

It is hardly surprising that the East Europeans would want NATO tripwire forces. Those American proponents of expansion who argue otherwise ignore the history of the alliance during the Cold War. One of the earliest and most consistent aspirations of the West European members of NATO was to establish and keep a U.S. troop presence on the Continent.[23] Without that presence, European leaders stated privately (and sometimes publicly), they could never be certain that the United States would honor its pledge to defend the European signatories if war actually broke out. Not only did the allied governments want U.S. forces stationed in Europe, they wanted them (along with nuclear weapons) deployed in forward positions so that they were certain to be caught up in the initial stages of a Warsaw Pact offensive. A persistent feature of the transatlantic relationship throughout the Cold War was the West European effort to deny U.S. policymakers the luxury of choice.[24]

Atlanticists in the United States aided and abetted the agenda of the West Europeans. In the so-called burden-sharing debates of the 1970s and 1980s, moderate critics of NATO argued that while it might be in America's best interest to remain in the alliance, the United States had no need to keep troops in Europe—that, in fact, such a high-profile U.S. presence enabled the allies to "free ride" on Washington's security guarantee. But Atlanticists scorned the notion that U.S. membership in NATO would be sufficient to help protect Western Europe's security. Even modest force reductions, as embodied in the Mansfield Amendment and similar proposals, they warned, would undermine NATO's effectiveness and risk nervous West European governments' drifting into de facto neutrality.[25] The troop presence was indispensable, they insisted, to reassure the allies and make the U.S. defense commitment credible.

Given that history, it is inconsistent, if not disingenuous, for some of those same Atlanticists to foster the impression that the East Europeans would regard NATO paper promises as sufficient. The West Europeans were unwilling to trust the U.S. treaty commitment to aid them, despite the importance to America's own security of keeping Western Europe out of Moscow's orbit. The East Europeans, who recognize that their region has never been as important to the United States, would have an even greater reason to want the tangible reassurance provided by a NATO tripwire that included U.S. troops.

It is true that no prominent American advocate of NATO expansion has yet argued for stationing U.S. troops in Eastern Europe, and on the surface that might appear to be an extremely remote possibility. But it is worthwhile to recall that the prospect of a permanent U.S. military garrison in Western Europe seemed equally unlikely when NATO was created in 1949. In fact, Secretary of State Acheson, Chairman of the Joint Chiefs of Staff Gen. Omar Bradley, and other officials of the Truman administration explicitly assured Congress and the American people that the United States would *not* station troops on the Continent.[26] Less than two years after the North Atlantic Treaty was safely ratified, however, some four U.S. divisions were on their way to Europe. More than four decades later, and despite the disintegration of the Soviet Union, they remain as a tangible symbol of the U.S. security commitment. A similar "follow-on" to NATO's expansion into Eastern Europe cannot be ruled out.

21

Supporters of enlargement do not address that point. Yet given their own arguments stressing the importance of stability in Eastern Europe, the need to discourage any future Russian expansionist bid, and the danger to American interests posed by conflicts in the region, that is a curious omission. They insist that the stakes are vital, but they imply that such important interests can be effectively secured by merely adding signatures to a piece of paper. Atlanticists would apparently like to convince the American people that while it is imperative to guard the security of Central and Eastern Europe, there will be no need for the United States to incur significant costs and risks. Perhaps they fear that there would be little domestic support for stationing American troops in such a volatile region—a perception that is probably correct—or perhaps they have not thought through all of the implications of their policy prescriptions. In any case, there is a conceptual chasm between the importance Atlanticists attach to the goals of stabilizing Central and Eastern Europe, and securing democracy there, and the means they publicly advocate for achieving those goals. Enlarging NATO without resolving that contradiction creates the prospect that the United States will drift into commitments, the implications of which are not fully comprehended and the dangers of which most Americans would not wish to embrace.

A "new" NATO may emerge from the incremental enlargement of security jurisdiction rather than from an increase in the roster of formal members—or increased jurisdiction may at least predate larger membership. Indeed, such a creeping enlargement of security responsibilities appears to have been under way even before the proliferation of prescriptions for NATO's expansion or the unveiling of the Partnership for Peace. As early as June 1991, the NATO foreign ministers issued a warning, clearly aimed at Moscow, that any attempt to undermine the newly won freedom and independence of the East European states would be a matter of "direct and material concern" to NATO.[27]

An even clearer indication that at least some officials in the United States were beginning to assume that the political welfare of Eastern Europe was a vital interest of the alliance came the following year. The initial draft of the Defense Department's planning guidance document for 1994–99, leaked to the press in February 1992, contained several "illustrative" war scenarios. One of the more interesting scenarios posited a U.S.-led NATO response to an invasion of Lithuania by Russia and Belarus. The authors of the planning guidance document described how such aggression would trigger a NATO counterattack ultimately involving more than 7 American Army divisions, 45 fighter squadrons, 4 heavy bomber squadrons, 6 aircraft carrier battle groups, and a Marine expeditionary force.[28] In other words, the Pentagon planners were operating on the assumption that NATO would respond to an attack on an East European country as it would to an attack on an alliance member.

That kind of thinking is exceedingly worrisome. Conceptually, NATO's security jurisdiction is already moving eastward, years before the first formal membership invitations are likely to be sent out. Those who recognize the potentially disastrous consequences of NATO's expansion may have more to fear from initiatives such as the Pentagon's planning document and the Partnership for Peace than they do from a formal motion to bring the East European nations into the fold as full-fledged members. NATO's expansion can be de facto as well as de jure, and the former is more likely in the short term.

### Partnership for Peace: Evasion or Covert Expansion?

The Clinton administration has found itself buffeted by political crosscurrents, both foreign and domestic, on the issue of NATO's future. That turmoil became especially pronounced in late 1993 and early 1994. A growing number of East European governments were lobbying hard for immediate (or at least prompt) inclusion in the alliance. Proponents of expansion in the United States, including several major past and present political figures, were also pressing for meaningful action on the expansion issue, although as noted, they differed sharply among themselves on the scope of expansion as well as the underlying strategic rationale.

The West European allies were badly split on the issue. Bonn was the most inclined to support membership for at least the Central European countries. As columnist Jim Hoagland observed, the reasons for that position were not difficult to fathom. "With the outcome of the Russian revolution still uncertain, Germany would prefer to push NATO's frontier to the east so that any fighting would be done on Polish soil, rather than on German territory. Getting Poland into NATO means Germany would cease being NATO's eastern frontier guard and would move to the center of Europe, psychologically and geographically."[29] Britain, France, and several other European NATO members, however, seemed noticeably reluctant to involve NATO in the volatile

23

security affairs of Eastern Europe. (In an unusually candid moment, German defense minister Volker Rühe conceded that his country and the United States were the only NATO members that were "enthusiastic" about expanding the alliance and that some signatories were distinctly cool toward the idea.)[30]

More troubling, after its initial passive response, Moscow expressed vehement opposition to any plan to enlarge NATO that did not specifically include Russia. That opposition was of particular concern to U.S. policymakers. Key administration leaders, especially Strobe Talbott, the president's principal adviser on Russian affairs—and now deputy secretary of state—believed that maintaining a harmonious relationship with a democratic Russia ought to be Washington's highest policy priority. They were worried that expanding NATO in the near future, over Moscow's explicit objections, would inevitably complicate if not poison U.S.-Russian relations. Even worse, such precipitous action threatened to undermine the already shaky Yeltsin government, with dangerous, unpredictable consequences. The Clinton administration's sensitivity to that danger was captured in the comments of Robert Hunter, the U.S. ambassador to NATO, who emphasized that "we don't want to put a cudgel in the hands of the enemies of reform in Russia."[31]

Finally, there was concern on the part of at least some U.S. policymakers that the expansion process could gain a momentum of its own and snowball out of control. The administration would probably have been willing to endorse relatively quick membership for Poland, Hungary, and the Czech Republic—assuming the problem of Russian opposition could be overcome—but it saw no easy way of halting enlargement, even temporarily, at that point. Hunter candidly expressed that apprehension. "If NATO were to take in Poland, what would we say to Ukraine? If we take in Hungary, what do we say to the Baltics?"[32]

The administration's response to those conflicting pressures was the Partnership for Peace. *Financial Times* columnist Edward Mortimer correctly identified the primary political motivation for the Partnership for Peace. "PFP, for all the grand rhetoric it has been wrapped in, is transparently a device for gaining time, for postponing invidious choices."[33] Nonetheless, the underlying logic of the program points in the direction of eventually enlarging the alliance.

Although the Partnership for Peace appeared to be primarily the brainchild of the Secretary of Defense Les Aspin, the most detailed exposition of the concept came in a speech by Secretary of State Christopher at the NATO foreign ministers' meeting in December 1993.[34] Christopher emphasized that the alliance envisioned "defense cooperation developing in a broad range of fields." The Partnership for Peace would provide a means for NATO's eastern neighbors to "develop a practical working relationship" with the alliance, and each state would determine what resources it was willing to commit to that relationship. Even at the outset, though, "the Allies should provide all participants with a pledge of consultation in the event of threats to their security." New partners that had once been part of the communist world were expected to make important changes as well, including adapting their defense structures to ensure reliable civilian control and making their military doctrines and weapons systems compatible with NATO's.

The Partnership for Peace was designed to be more than a paper relationship, Christopher stressed.

> It will develop capabilities to meet contingencies, including crisis management, humanitarian missions and peace-keeping. It will develop useful habits of cooperation. It will enable us to develop common military standards and procedures. Peace partners will train side-by-side with NATO members and take part in joint exercises.[35]

Christopher's remarks exhibited "on the one hand, on the other hand" characteristics throughout, reflecting the administration's urgent wish to satisfy everyone. "We must help to fill the vacuum of insecurity and instability that has come with the demise of the Soviet empire," he intoned. "We must build the structures and the patterns of cooperation that will help to ensure the success of democracy and free markets in the East." Noting that NATO leaders had created the North Atlantic Cooperation Council two years earlier, Christopher added: "With the Partnership for Peace, we can deepen NATO's engagement with the East. We must demonstrate that the West is committed to helping Europe's new democracies address some of their most immediate security problems." Almost immediately, however, he tempered those bold pronouncements. "At the same time, we should signal an evolutionary transformation of the alliance." How one addresses "immediate" security problems with an "evolutionary" transformation, he did not say.

The same effort to reconcile different perspectives was evident in other comments. The Partnership for Peace "will be a military relationship," but it will also "have a strong political dimension." The new partners "should finance their own involvement," but "some new NATO resources will be necessary." The foreign ministers should make clear that "as a matter of principle, NATO is open to new members." On the other hand, "many factors will enter into decisions about expanding NATO membership."

Christopher's presentation had all the characteristics of a policy designed by a committee. Perhaps more accurately, it was the product of an effort to satisfy an assortment of conflicting domestic and foreign political and diplomatic constituencies. Nevertheless, those who favored ultimate expansion of the alliance had more reason to take heart from his remarks than did those who opposed it.

In addition to the substantive steps of greater military cooperation outlined in the speech, Christopher made it clear that the Partnership for Peace was the preparatory stage for the enlargement of NATO's responsibilities and, in all probability, the roster of members. Although the new arrangement was "an important step in its own right," it could also "be a key step toward NATO membership." The secretary of state reminded his audience of the immediate and long-term implications: "The Alliance must understand that this Partnership represents a decisive commitment to become more fully engaged in security to the East."

That was indeed the bottom line. Advocates of immediate expansion are not content with the Partnership for Peace, of course. Gary Geipel's response typifies the reaction of that camp; he contended that the "Partnership for Peace is not good enough." The policy "does not alleviate Central Europe's basic insecurity, and therefore it does nothing to insulate existing NATO members from potential instability on the eastern border of the alliance." At the very least, Geipel argues, "explicit criteria and conditional timetables for the admission of Central European members" should have been established.[36]

Geipel and other boosters of immediate expansion are unlikely to be disappointed in the long run. In addition to the sentiments expressed by Christopher and other U.S. officials, the text of the invitation extended by the heads of state and government at the January NATO summit states, "Active participation in the Partnership for Peace will play an important role in the evolutionary process of the expansion of NATO."[37] Despite all its evasions and qualifiers, the Partnership for Peace is the first step on a very slippery slope leading to the eastern extension of NATO's security jurisdiction.[38]

It is also a formulation that fails to address some of the more troublesome and hotly contested issues. Most crucial, the Partnership for Peace still begs the question of whether Russia will be included or excluded at the time enlargement occurs. It is possible that such ambiguity may simply reflect the conceptual confusion that has characterized other elements of the Clinton administration's foreign policy. On the other hand, the evasion may be a clever ploy by the administration to postpone the day of decision on what is certain to be a difficult and potentially explosive matter.

As now constituted, the Partnership for Peace is sufficiently flexible that it could lead either to a gradual expansion of NATO to constrain Russia or to a new, all-inclusive NATO with a pan-European security focus. An important factor might well be Washington's assessment of the extent and durability of democratic political reforms in Russia as well as Moscow's foreign policy orientation. Although the administration seems to be leaning cautiously toward an inclusive policy, there are occasional hints that U.S. policymakers are keeping open the option of transforming the Partnership for Peace into an arrangement to contain Russian power, if the need arises.[39]

The Clinton administration and other cautious proponents of expansion may wish to disguise their ultimate goal so as not to alarm the already suspicious Russians as well as those individuals and groups in the United States and Western Europe who are uneasy about the implications of NATO's enlargement. But the administration's disagreement with the more aggressive supporters of expansion is largely about timing. Too often the debate in the United States over NATO policy takes the sterile formulation of "expansion now or expansion later." In the final analysis, the Partnership for Peace serves as a rhetorical sedative to lull opponents of NATO's expansion into complacency. That could be seriously detrimental to America's well-being, for the expansion of NATO is unsafe at any speed.

# 2. NATO's Cold War Rationale and America's European Interests

There was little doubt that NATO was created primarily to deter Soviet intimidation of or aggression against Western Europe. True, the text of the North Atlantic Treaty does not mention the Soviet Union, and both the European and the American architects of the alliance were usually careful not to portray it publicly as an anti-Soviet organization. Most proponents maintained that the treaty was not directed against any specific nation but merely against "aggression." Sen. Tom Connally (D-Tex.), chairman of the Senate Foreign Relations Committee, typified the arguments of NATO supporters when he asked, "Who has any right to fear this treaty unless he is an aggressor or intends to become an aggressor?" Such reasoning failed to impress Moscow, which complained loudly and frequently throughout the treaty debate in 1949 that NATO was an aggressive, anti-Soviet alliance. Secretary of State Dean Acheson responded to such complaints with the biblical phrase, "The guilty flee when no man pursueth."[1]

There were important diplomatic reasons for that ploy, most notably the need to make the new regional defense arrangement consistent with the principles and objectives of the United Nations, but few policymakers took such rhetoric seriously. Senate hearings (especially the hearings conducted in executive session) were filled with statements by legislators and executive branch officials that the alliance was needed to deter the Soviet Union from pursuing expansionist designs against Western Europe. President Harry S Truman's speech following the signing ceremony for the North Atlantic Treaty in April 1949 also hinted none too subtly at that motive. Truman insisted that if such a treaty had existed in 1914 and 1939, "it would have prevented the acts of aggression which led to two world wars." He likened the alliance to "a group of householders, living in the same locality, who decided to express their community of interest by entering into a formal association for their mutual self-protection."[2]

Two things were notable about Truman's statement—and indeed most of the public statements in favor of NATO. First, the motive cited for the alliance was the prevention of another large-scale war in Europe. The benefit of NATO, as the president indicated, would be to prevent tragedies similar to those that erupted in 1914 and 1939. Second, that focus implied a mission to deter a major aggressor. In the context of 1949, there was only one nation in Europe capable of waging a war of conquest comparable to Hitler's rampage or even the more limited expansionist drive of Wilhelmine Germany in World War I. That country was the Soviet Union. Truman knew it, his advisers knew it, and so did everyone who heard his speech.[3] Although NATO might also have the salutary effect of preventing the resurgence of a revanchist Germany in the long term—or even the unexpected emergence of another dangerous expansionist power at some point in the future—the focus at NATO's founding was on one potential source of aggression: the USSR. Two decades after NATO's founding, President Richard M. Nixon bluntly expressed the consensus view in an address to the North Atlantic Council: "NATO was brought into being by the threat from the Soviet Union."[4]

True, some U.S. policymakers may have had other purposes in mind when they proposed NATO and other components of Washington's post–World War II security architecture. RAND Corporation analyst Benjamin Schwarz contends that containing Soviet power was only one reason U.S. leaders thought it necessary to "secure" Western Europe. Their broader objective, Schwarz argues, was to build an international political and economic order based upon "a preponderance of [U.S.] power."[5] That goal required integrating Germany and other West European states into a network of political, economic, and security ties. NATO was the key institution in that strategy.

The authors of NSC-68, the National Security Council's 1950 blueprint for Washington's Cold War strategy, insisted that the objective of building a stable international order ruled by a preponderance of American power was independent of any actions the Soviet Union might take. It was a policy, the authors conceded, "which we would probably pursue even if there were no Soviet threat."

But as Schwarz and coauthor Christopher Layne point out in *Foreign Policy*, although the Soviet threat was not central to the conceptual underpinnings of Washington's post–World War II policy, the Soviet Union's existence was "indispensable to that policy's success."[6] The reason for the apparent paradox was simple: the American people would probably not have supported a risky, expensive, long-term U.S. military commitment to Europe without the perception that an extremely serious threat to America's security was located on the Continent. In a very real sense, the Soviet Union—or something similarly scary—was a necessary enemy.

NATO's anti-Soviet focus became increasingly pronounced as the Cold War deepened. The memoirs of presidents and other U.S. policymakers who served during that period establish that connection beyond dispute. Their accounts of alliance affairs invariably stressed the need to protect Western Europe from Soviet aggression, described Moscow's repeated efforts—aided and abetted by Soviet sympathizers in Western Europe—to undermine the alliance's cohesion and resolve, and expressed the belief that NATO was indispensable as long as the Soviet Union continued to exist as a heavily armed totalitarian state. If NATO had other purposes during the Cold War, two generations of U.S. leaders inexplicably failed to emphasize them in their public statements. It was only when the Cold War started to wane that some policymakers began to suggest that the alliance's mission was much broader than preventing the Soviet hordes from overrunning democratic Europe.

Despite the looming Soviet threat at the time of NATO's birth, Congress and the American people initially sought to keep Washington's role in the alliance relatively modest. Truman administration officials, for example, assured senators as well as the public that the United States would not be expected to station troops on the Continent. They portrayed NATO as an alliance in which the West Europeans would have primary responsibility for the defense of their region, with the United States merely supporting their efforts. W. Averell Harriman, Washington's Marshall Plan representative in Europe, stated that the alliance was intended to establish "a balance of power in Europe, *backed up*, of course by the military establishment that the United States will have."[7] Officials certainly did not imply that the United States would have the primary responsibility for the defense of Western Europe—much less that the West Europeans would remain forever dependent on U.S. protection.

Less than two years after the creation of NATO, Washington's responsibilities increased markedly. But they did so because of an improvised response to extraordinarily threatening global conditions, not because of a new, long-term grand strategic design. Three factors converged to produce additional U.S. obligations: the unexpectedly slow pace of Western Europe's rearmament, a growing concern among American officials that the West Europeans would be unable to build an adequate conventional defense force without the addition of West German manpower—which France and several other allies opposed—and the fear that the Korean conflict might be merely the precursor of a global communist offensive at a time when the Europeans were not yet capable of defending themselves.

Those fears (combined with self-serving lobbying on the part of the allied governments) caused Washington to take two important steps. The United States agreed to station U.S. air, naval, and ground units in Europe as part of a combined NATO defense force, violating promises that had been made to the Senate at the time the North Atlantic Treaty was ratified. Washington also agreed to have a U.S. officer—Gen. Dwight D. Eisenhower—command the new NATO force. Administration leaders stressed that the assumption of additional responsibilities was not intended to be permanent, that the United States would bear them only until the West Europeans could complete their rearmament efforts. The allies had different ideas, however, and thus began the process by which the United States would come to have primary responsibility for Western Europe's defense instead of merely backstopping West European efforts.[8] Moreover, the "temporary" additional obligations have continued for nearly four and a half decades and show no signs of ending.

The circumstances surrounding NATO's birth are not merely a matter of historical interest. Both the emphasis on the security threat posed by a supposed superpower and the assurances that America's commitments to Western Europe were going to be limited were essential in gaining congressional and public approval of the alliance. The ratification debate might have turned out quite differently if NATO's sponsors had told the Senate and the American people that the United States proposed to subsidize Western Europe's defense indefinitely—even after the region recovered fully from the economic devastation wrought by World War II—that America intended to treat the region as a pampered protectorate even if the Soviet threat disappeared; and that American troops might eventually be involved in trying to settle every armed conflict on the Continent, no matter how obscure and parochial.

Undertaking risky and expensive obligations to keep a temporarily weakened Western Europe out of Moscow's orbit seemed to most Americans to be important to America's own security. But in the mid-1990s, neither consideration applies. The current generation of U.S. policymakers seems unable to distinguish between legitimate American interests in Europe and a needless (and unprofitable) attempt to act as the Continent's military nanny.

## The Nature of America's European Interests

The United States does have some legitimate interests in Europe, but they are relatively narrow and specific. Atlanticists habitually fail to understand that not every unpleasant development that might occur somewhere in Europe ought to be a matter of grave concern to American policymakers. Whenever critics of an excessively interventionist policy attempt to distinguish between vital and peripheral matters, NATO proponents inevitably respond with cries of "isolationism." Even a reasonably sophisticated scholar such as University of Chicago professor Charles L. Glaser succumbs to that temptation. Citing the writings of Christopher Layne, Earl C. Ravenal, and the late Eric Nordlinger, Glaser contends that such "isolationists" believe, "now more than ever, that whatever dangers might threaten Europe will not threaten the United States." He responds that despite the end of the Soviet threat, U.S. security has not been "entirely separated from the future of Western Europe." Because the United States could still be drawn into a major European war, Washington should not be "unconcerned" about developments in Europe.[9]

But Glaser and those who make similar arguments merely erect a straw man and then proceed to pummel it. The authors Glaser cites would not make the categorical statement that "whatever dangers" might emerge in Europe would be immaterial to the United States. Layne, in particular, has stated repeatedly that America does have some interests in Europe and cannot be oblivious to events there. He merely argues that bona fide U.S. interests on the Continent are not nearly as extensive as Atlanticists insist. Although other foreign policy writers can—and do—disagree with that conclusion, Layne's position is a far cry from the caricature of an isolationist, fortress America mentality. It is hardly isolationism to suggest that many, or even most, disturbances in Europe will not affect America's security. The pertinent issues are how to define America's European interests—separating relevant from irrelevant matters—in a way that makes sense in the post–Cold War setting and how to defend and promote those interests in the most low-risk, cost-effective manner possible. Dismissing nuanced criticism of Washington's existing European policy as isolationism does not contribute to the formulation of a coherent post–Cold War strategy.

Nor is it helpful to make sweeping statements that "Europe" is a vital interest of the United States and to imply that Washington must be deeply concerned about all developments on the Continent. National Security Adviser Anthony Lake exhibits such fallacious reasoning when he states, "If there is one thing this century teaches us, it is that America cannot ignore conflicts in Europe."[10] Lake misses the crucial point that both of the conflicts in which the United States ultimately intervened were wars involving all of Europe's great powers. Such serious disruptions of the international system obviously had high potential to place important American security interests at risk. (Although in the case of World War I, the United States confused a shift in the European balance of power—one that Washington could have tolerated, albeit with some discomfort—with a complete breakdown that might well have posed a dire threat to America's security.)

Not every conflict that has erupted in Europe, or is likely to in the future, necessarily has wider strategic implications. That is particularly true if the war does not involve rival alliances of European great powers. There is no validity to the notion that more limited struggles, especially those involving small powers in Eastern Europe, are destined to escalate to continent-wide conflagrations that will automatically drag in the United States.

In fact, there was nothing "inevitable" about the U.S. entry into the two world wars. The United States could have remained aloof from World War I if President Woodrow Wilson and his advisers had not overrated the severity of Germany's potential threat to U.S. security (and if Wilson himself had not been such a rigid moralist who was willing to take his country into a bloody conflict to uphold the sacred right of American citizens to travel on the ships of belligerent nations). Indeed, given the disastrous aftermath of World War I, both America and Europe might have been better off if the United States had not intervened to thwart a victory by Wilhelmine Germany.

There is even a plausible, albeit much weaker, argument that the United States could have chosen to remain on the fringes of World War

II instead of becoming a belligerent.[11] Yale University political scientist Bruce Russett presented that case more than two decades ago, arguing that Britain and Russia could probably have held Germany at bay without the direct involvement of the United States. The most likely outcome of such a conflict, he concluded, would have been an exhausting military stalemate between Nazi Germany and the Soviet Union and, perhaps, the ultimate overthrow of both totalitarian regimes.[12] Although Russett's thesis is plausible, nonintervention would have been an extraordinarily high-risk strategy for the United States to adopt. Some risks are, perhaps, better not taken. Nevertheless, it is important to remember that U.S. belligerency even in that most terrible war was the result of policy calculations by U.S. officials, not of some inexorable geopolitical dynamic.

It is instructive that, before the two great wars of the 20th century, there were numerous conflicts in Europe that did not lead to U.S. involvement. And one does not have to go back to the 18th or early 19th century for examples; such conflicts also occurred in the late 19th and early 20th centuries. They included the Prussian-Austrian War of 1866, the Franco-Prussian War of 1871, the First Balkan War (1912), and the Second Balkan War (1913).

Those who contend that the United States must now be deeply involved in Europe's security arrangements invariably argue that the revolution in technology, especially military technology, during the 20th century has erased the significance of distance, thereby making it impossible for America to avoid European conflicts. That thesis, although it has some validity, is too facile. There are two more plausible reasons that the United States was able to remain aloof from those earlier conflicts but did not stay out of the two world wars. The first factor was the influence of an entrenched policy, which began with Washington and Jefferson, that emphasized the need to avoid becoming entangled in European conflicts whenever possible. The second, equally important reason is that none of the earlier conflicts involved an effort on the part of a major state to completely overturn the European balance of power. U.S. policymakers, therefore, correctly concluded that those wars did not pose a threat to America's well-being.

That is a pertinent lesson for the post–Cold War era, because there is no inherent reason why political turbulence or even the occasional outbreak of war in Eastern Europe should threaten America's security any more than did, for example, the Franco-Prussian War. Several variables determine whether a conflict is merely of regional importance or has wider implications that might impinge on U.S. interests. Those variables include the location and scope of the armed struggle; the number of nations involved; and especially the size, influence, and power of the key belligerent states. Some conflicts might warrant a U.S. response, others would not. The argument that the United States must be concerned about all European conflicts is not justified historically or logically.

The notion that "Europe" is a vital—or at least an important—interest of the United States leads to the wrong analytical approach. Europe (or any other region) per se does not constitute an American security interest. Rather, the United States may have an important interest in promoting certain beneficial developments and preventing other, adverse developments in a region. That is a subtle but important distinction that implies greater selectivity and precision.

The core U.S. interest in Europe is to prevent any power or combination of hostile powers from achieving a hegemonic position and thereby controlling the major industrial states of Western Europe. Such a massive disruption of the European balance of power could pose a serious threat to America's security. The population and technological resources of Western Europe make that region a rich prize for large expansionist powers. A major motive for the U.S. entry into both world wars as well as Washington's willingness to become a member of NATO was to prevent Western Europe from being at the disposal of such an expansionist state.

No hegemonic threat exists today, nor is there a credible one on the horizon. Russia has monumental economic and political woes that will undoubtedly keep it occupied for years. Its annual military spending has already declined to $47.22 billion—only modestly more than the military budgets of Britain or France—and it may decline to as little as $20 billion in the coming year.[13] The readiness of Russia's armed forces continues to erode as well. The *Wall Street Journal*'s Therese Raphael notes the "very low morale in the Russian army. Much of it is incapacitated by lack of supplies and demoralized by internal divisions. This state of disintegration has left the army nearly incapable of acting as a coordinated military unit." She quotes an estimate by Sergei Stepashin, former chairman of the Russian parliament's Defense

Committee, that "the rapid hemorrhaging of soldiers has left a top-heavy army with 630,000 officers commanding only 544,000 men."[14]

The Russian navy is not in much better shape. In February 1994 naval leaders announced that they were mothballing three of Russia's five aircraft carriers. All three needed to be overhauled, the officials stated, but there were no shipyard facilities in operation that could do the work, nor was there any money to pay for it.[15] The fact that the carriers were relatively new—they had all been built in the late 1970s or early 1980s—and could normally have remained in service for decades highlighted the drastic nature of the decision and the financial crisis that precipitated it.

That is not to say that Russia will never seek to reemerge as one of the world's great powers. It is in fact already making limited efforts to do so, primarily by attempting to contain (and perhaps sometimes manipulate) conflicts in its "near abroad," that is, the newly independent republics on its borders. But there is no evidence that Moscow will be able to pursue more extensive ambitions in the foreseeable future. Certainly, Russia does not pose a serious conventional military threat to Western Europe.

There is even less to fear from the only other credible candidate for European hegemony—Germany. To be sure, Bonn will play an increasingly assertive role in world affairs. Bonn's growing use of economic leverage in the European Union, as well as in Eastern Europe, and its declared desire for a permanent seat on the UN Security Council are early indications of that trend. Nevertheless, with a defense budget of barely $29 billion a year, Germany is far from becoming a military powerhouse. Moreover, German military spending and force levels are declining, not rising—a curious strategy for any nation with expansionist ambitions.

Whatever the accuracy of the judgments made by U.S. policymakers in the cases of the two world wars and the Cold War, there is at least a colorable American security rationale for preventing a hegemonic challenger from gaining a valuable geopolitical objective. A smothering strategy—attempting to pacify every portion of the Continent, however remote and obscure, and resolve every ethnic feud or territorial dispute that might lead to armed conflict—is quite another matter. Such a goal is unnecessary as well as unattainable.

In that context, it is necessary to distinguish between the relevance of Western Europe and Eastern Europe from the standpoint of America's interests. Vice President Al Gore expressed the new conventional wisdom when he contended, "The security of the states that lie between Western Europe and Russia affects the security of America."[16] That statement and others like it are without historical foundation. Eastern Europe has never figured prominently in this country's strategic calculations. Keeping Western Europe out of hostile hands is an important (although not, as overwrought Atlanticists typically argue, a vital) American security interest.[17] The political status of Eastern Europe simply does not have the same relevance. The ability of the United States to tolerate Soviet domination of the region throughout the Cold War confirms that point.

Although Americans sympathized with the plight of East European populations who suffered under Soviet imperialism and rejoiced when they were able to secure their independence, few serious U.S. policymakers even contemplated military action to dislodge the Soviets. Occasional suggestions for a "rollback" strategy were summarily dismissed by liberal and conservative administrations alike. The refusal to incur the costs and risks of securing a friendly Eastern Europe contrasted sharply with the willingness to assume significant burdens and dangers decade after decade to keep Western Europe out of Moscow's orbit. The burden of proof is on those who contend that the independence and stability of Eastern Europe have now become so important that the United States must risk being entangled in the conflicts of that region through an expanded NATO.

Yet that is the import of the comments of such officials as Secretary of State Warren Christopher. Meeting with the foreign ministers of nine Central and East European countries in July 1994, Christopher assured them that the security of their region was of vital interest to the United States. "Peace and stability in Europe require that the nations of Central and Eastern Europe be strong, independent, democratic and secure," he stated in opening comments to the group. "The United States has a vital interest in ensuring that goal is achieved." Placing additional emphasis on that theme, Christopher stated, "For the United States there must not and cannot be a so-called gray zone of instability in Central and Eastern Europe."[18]

Christopher's comments illustrate the tendency of establishment foreign policy figures to use the term "vital interests" carelessly and promiscuously. That is a habit that policymakers must break. Declaring something a vital interest implies a willingness on the part of the

United States to risk a major war to defend it. Otherwise, the term is simply a fatuous rhetorical flourish. If Christopher is to be taken at his word, Washington must be willing to incur the risk of war not only to keep Eastern Europe out of Moscow's sphere of influence but to oust indigenous undemocratic regimes and suppress internecine struggles as well as conflicts between states, however small, anywhere in the region. That would constitute an unprecedented expansion of U.S. responsibilities in Europe for which there is little evidence of domestic support and no compelling justification.

Leading the alliance on a mission to establish and maintain political stability throughout Central and Eastern Europe would constitute a foolish attempt to micromanage the Continent's security affairs. In the absence of a great-power challenger that could dominate the Continent and thereby increase the potential threat to the United States, Americans can afford to view lesser regional and internal conflicts with considerable detachment. Only if turmoil in Eastern Europe and the states of the former Soviet Union reached the point that it threatened to engulf Western Europe, and the West Europeans proved incapable of containing the threat, would an arguable case exist for U.S. action. But contrary to alarmist scenarios—replete with images of toppling geopolitical dominoes—the conflicts taking place in Eastern Europe, however numerous, range from the petty to the parochial. They may be important to the parties directly involved, but virtually none of them has the potential to disrupt even the European, much less the global, balance of power. Those disorders do not reach (indeed, in most cases they do not even approach) the threshold at which U.S. military involvement might be warranted.

Moreover, there is no reason to assume that the prosperous and capable powers of Western Europe would remain catatonic while conflicts to their east threatened their own vital interests. If disorders among their eastern neighbors escalated to the point that they threatened to spill over, there is every reason to expect that the West European countries, as rational and competent security actors, would take steps to counter that danger. The appropriate mechanism for containing, and when possible resolving, East European conflicts is not a meddlesome, enlarged NATO led by the United States but more tightly focused, Europeans-only security arrangements.

## The "Interdependence" Argument

American proponents of NATO's expansion generally fail to formulate a coherent definition of American interests in Europe. They show even less ability to comprehend that some portions of Europe are more important than others to the United States—or that some adverse developments are more important than others. To die-hard Atlanticists, Europe is a kind of geopolitical puzzle in which all of the pieces, however small, are interconnected and vital to the whole.

Sen. Richard Lugar's (R-Ind.) thesis is typical of such reasoning. The foundation of his case is that a prosperous Western Europe is crucial to America's own economic well-being. "The U.S. cannot allow Europe to unravel for the third time in this century," Lugar stresses. "The U.S. will not be able to pursue domestic economic renewal and reconstruction successfully without peace and stability in Europe." Prosperity at home, he concludes, "requires a more involved and engaged approach abroad."[19]

That is essentially a "lightswitch" theory of stability—with only two positions: off or on. To Lugar and those who share his views, there is no possible middle ground. European stability means the exceptional ultrastability that has characterized Western Europe since the end of World War II. Anything less signifies impending chaos and a cascading effect that would culminate in the "loss" of Europe and economic as well as strategic calamity for the United States.

Lugar leaves little doubt that his policy assumptions are based on such a nightmare scenario. "The problem is not just the tragedy of Bosnia or even that Eastern Europe as a whole might unravel. The danger is that Europe as a whole could again come apart." And only the expansion of NATO into Eastern Europe can prevent such a tragedy.

> The critical strategic change in the new Europe is that the locus of conflict has shifted from its center to the periphery, or the so-called twin arcs of crisis to the east and south. However, potential and real crises on the periphery are not peripheral to the future of European stability. There can be no lasting security at the center without security at the periphery.[20]

The premise that every development in Europe is inextricably connected to every other development leads Lugar to make some extraordinary claims. For example, the Balkans are not merely another post–Cold War geopolitical backwater characterized by the same kinds of ugly tribal conflicts that have erupted in so many other regions. Instead, "The Balkans—the fault line between east and west since Roman-Byzantine days—is the strategic heart of the world."[21]

It would be tempting to dismiss such hyperbole with a chuckle— except that it is being expressed by one of the most influential members of the U.S. Senate on foreign policy matters. Moreover, it is the implicit or explicit assumption of many Americans who advocate NATO's military intervention in the former Yugoslavia and, more generally, the adoption of the pacification of Eastern Europe as NATO's new mission.

Benjamin Schwarz identifies the central problem with Lugar's line of reasoning.

> The logic that dictates an expansion of NATO's responsibilities eastward has extremely unsettling implications. . . . After all, if the United States, through NATO, must guard against internal instability and interstate security competition not only in Western Europe, but in areas that could infect Western Europe, where would NATO's responsibilities end? It is often argued, for instance, that the alliance must expand eastward because turmoil in East Central Europe could provoke mass immigration flows into Western Europe, threatening political stability there. Of course, turmoil in, for example, Russia or North Africa could have the same effect, as could instability in Central Asia (which could spread to Turkey, spurring a new wave of immigration to the West). Must NATO, then, expand even further eastward and southward than is currently proposed?[22]

The logic of Lugar and other NATO expansionists indeed points inexorably in that direction. Advocates of comprehensive enlargement clearly embrace the proposition, since the inclusion of Russia and other states of the former Soviet Union would take NATO not only into Eastern Europe but deep into Central Asia and even East Asia. As those countries sign on to the Partnership for Peace, NATO's security concerns will be moving far afield from Western Europe, albeit with less intensity than would be the case with the incorporation of additional alliance members.

But even proponents of more limited NATO expansion would find it difficult to halt the enlargement process at arbitrary points—at least in terms of security jurisdiction, if not formal membership. Lugar's reasoning implies that international stability and security are seamless webs—that one cannot be concerned about the security of Western Europe without also protecting the security of the "periphery." The further implication is that "Western Europe" itself is an artificial concept that may have had some meaning in the Cold War era, when the Iron Curtain sharply divided the Continent into Western and Soviet spheres of influence, but has little meaning in the far more fluid post–Cold War setting. If that is true, however, the concept of "Europe" is equally arbitrary from a security standpoint. As Schwarz points out, events outside Europe—in Central Asia, North Africa (and for that matter, the Middle East)—could also have an impact on European states. An attempt to stabilize Eastern Europe would inherently tend to draw NATO into the myriad conflicts of those other regions, in a veritable daisy chain of security commitments that would take the alliance ever farther from the "center" in Western Europe.

Lugar and his ideological colleagues might well recoil from the prospect of such an amorphous geographical jurisdiction for NATO, but they must then reassess their entire line of reasoning. If it is possible to limit U.S. security concerns to all or part of Eastern Europe without having to address the quarrels and conflicts of neighboring regions, it is also possible to limit U.S. concerns to the security of Western Europe without having to address the quarrels and conflicts of Eastern Europe. Conversely, if it is impossible to do the latter, there is no logical reason why it should be possible to do the former. Lugar and the other would-be limited expansionists cannot have it both ways.

Distinguishing legitimate from illusory American interests in Europe raises serious doubts about the wisdom of maintaining U.S. membership in NATO even if the alliance confines itself to its traditional, Cold War era security role and geographic coverage. After World War II the United States became a de facto European power; indeed, it became the dominant player in the European state system. But the circumstances that led to that unprecedented role were exceptional. World War II had not merely disrupted—it had virtually obliterated—the European balance of power. Much of Western Europe lay in ruins; Britain and France, the traditional major powers in the region, were badly weakened and demoralized; and Russian military forces had penetrated farther west than at any time since the defeat of Napoleon. Germany had been eliminated as a political and military factor—meaning that there was no significant counterweight to Moscow's influence in Central Europe. The unbalancing of the European state system, which began with the disintegration of the Austro-Hungarian empire nearly three decades earlier, had reached its zenith.

Washington's choice was to move into the enormous power vacuum that the war had created or to concede that the Soviet Union would become the preeminent power in Europe—indeed, in all likelihood would emerge as the Continent's undisputed hegemonic power. Even if there had been a relatively benign regime in Moscow, U.S. policymakers would not have relished the prospect of one nation's having such a dominant position in Europe. That the likely hegemon was governed by an aggressive totalitarian regime made that option utterly unpalatable. The United States, therefore, became the geostrategic counterweight to the Soviet Union, since no European country or combination of countries was able to play that role.

Those conditions no longer apply. Today's Europe is fundamentally different from the post–World War II version threatened by Soviet domination. There is no obvious preeminent power, much less a would-be regional hegemonic state. Just as the post–Cold War European balance of power more closely resembles the structure that existed throughout the 19th century, with the existence of multiple centers of influence, so too does the threat environment approximate that of the earlier era. There will undoubtedly be disorder, confrontations, and occasional episodes of violent change, but there is no credible evidence that they will pose a threat to America's security even remotely comparable to the threats posed by the expansionist totalitarian powers during World War II and the Cold War.

Under such radically changed conditions, there is no longer a need for the United States to play the role of a de facto European power, with all the responsibilities that status implies. Although Washington can and should take some interest in European security matters, that level of concern is far different from seeking to remain the leading power on the Continent. Yet NATO, even in its present form, is based on the premise of continued U.S. preeminence.

A strategy based on protecting America's limited European interests does not require an activist, high-profile U.S. role in European affairs. Perhaps more important, it leaves no room at all for leading an expanded version of NATO on a quixotic adventure in Eastern Europe.

# 3. NATO Expansion: Playing Russian Roulette

Enlarging NATO is a faulty idea for several reasons. It would constitute a needless provocation of Russia, entangle the United States and the West European powers in numerous, dimly understood disputes in an extremely turbulent region, and exacerbate the already unhealthy dependence of the Europeans on the United States for their security. Any one of those factors would be reason enough to view the various expansion proposals with great skepticism. Taken together, they should constitute a flashing red light for U.S. leaders.

It would be extraordinarily difficult to expand NATO eastward without that action's being viewed by Russia as unfriendly. Even the most modest schemes would bring the alliance to the borders of the old Soviet Union. Some of the more ambitious versions would have the alliance virtually surround the Russian Federation itself. The only feasible way to avoid antagonizing Moscow would be to include Russia in the roster of new members, but that would merely trade one set of problems for another, equally dangerous set.

Most American expansionists contend that an enlargement of NATO is not motivated by hostility toward Russia. They argue that a new, larger NATO would be inclusionary rather than exclusionary—a "bridge" from Western Europe to Russia, not a barrier to protect the rest of Europe from Russia. That is generally true even of supporters of expansion who believe that Russia should not be included in the ranks of new members.[1] Ronald D. Asmus, Richard L. Kugler, and F. Stephen Larrabee typify that view when they state, "Extending the alliance eastward should be seen as the West taking a step toward Russia, rather than against it."[2] Moreover, many advocates of expansion contend that NATO is increasingly a political organization for fostering pan-European security cooperation and strengthening democratic political systems, not a military alliance in the traditional sense. Therefore, it should pose no menace to Russia; indeed a larger alliance would actually provide indirect benefits to Russia by stabilizing the zone of turbulence on its western frontier.

*Washington Post* columnist Stephen S. Rosenfeld offers perhaps the most extreme version of the thesis that the enlargement of NATO would be a benign act that would actually improve relations with Moscow.

> It patronizes Russians to think they cannot bring themselves to understand that NATO expansion (1) does not threaten them and (2) promises them security advantages and much more. By steadying a disruption-prone slice of Europe on a sensitive Russian border, expansion gives heart and political space to Russia's liberal Westernizing party and steals a card from the conservative and nationalistic party that is given to tension and adventure.[3]

Former ambassador Max M. Kampelman offers a similar view. He argues that the reluctance of the members of NATO to grant admission to Poland and the other Central European countries, "which has been justified by a fear of antagonizing Russia's military and nationalist fringe," is short-sighted. Such timidity "in the face of expansionist hysteria," he warns, "may prove counterproductive."[4]

## Growing Russian Animosity

A majority of enlargement enthusiasts may even believe their own conciliatory rhetoric, but expansion is likely to be viewed differently in Moscow—and by virtually all political factions, not just strident nationalists. Even when Russian president Boris Yeltsin briefly seemed to acquiesce to proposals to incorporate Poland and other Central European states into NATO, Russian defense minister Pavel Grachev took the extraordinary step of publicly disputing his president's statement. Yeltsin sharply rebuked such insubordination, asking, "Who is president of Russia, you or me?" but Grachev's views ultimately prevailed. In a September 30, 1993, letter to Western leaders, Yeltsin explicitly retracted his earlier statement, and the Russian government is now officially on record as objecting to an eastward expansion of NATO—especially if it excludes Russia.[5] Even if Yeltsin had not needed to call on the military to support him in suppressing the attempted coup by hard-line elements in parliament, it is unlikely that he could have persisted in a policy of appeasement toward the West on an issue that is so central to Russia's security interests. The debt he incurred to the military hierarchy for its support in the October 1993 crisis, however, eliminated even the slightest possibility of continued accommodation.

Since then, Russian statements concerning an enlarged NATO have become steadily more hostile. In late November, Yevgeny M. Primakov, director of Russia's foreign intelligence service, warned that enlarging NATO might lead to military countermeasures, including "a fundamental reappraisal of all defense concepts on our side, a redeployment of armed forces, and a change in operational plans." He added that expansion "will be taken by a considerable part of Russian society as the approach of danger to the Motherland's borders," and thus promote anti-Western sentiment.[6]

Even Foreign Minister Andrei Kozyrev and other moderates insist that if an expansion of NATO takes place, it must include Russia. Any other form of enlargement would be interpreted as an unfriendly act. As an interim measure, Russian leaders are also seeking a higher status for Russia within the Partnership for Peace structure. That special partnership would recognize that Russia is a nuclear superpower. Despite some expressions of flexibility, however, the United States and its European allies deflected Moscow's bid for an enhanced position within the Partnership for Peace system in the spring of 1994, offering only vague assurances that the alliance's relationship with Russia would be "broader" than the relationship with the other members of the partnership.[7] Although Yeltsin then indicated that Russia intended to join the Partnership for Peace, he added that Moscow would still put forward a "more comprehensive plan" to define its relationship with NATO.[8] Russia did in fact join the Partnership for Peace in June, with the NATO states acknowledging informally that Russia would be treated in an manner commensurate with its status as a superpower with nuclear weapons.[9] It was a telling development, however, that the United States and its allies were unwilling to sign a formal agreement spelling out such a special relationship or to give Russia a veto over the admission of new NATO members.

. The pervasiveness of Russian hostility toward an expanded NATO that excludes Russia can be gauged by the comments of Kozyrev, easily the most pro-Western member of Yeltsin's foreign policy and national security policy teams. On the eve of the January 1994 NATO summit, the Russian foreign minister presented his government's case in the pages of *Frankfurter Rundschau.* Russia "has no right to stipulate who may and who may not join NATO," Kozyrev conceded, but it does have the right to insist on the principle, "enshrined in NATO's articles, . . . that NATO remains open to any democratic European or Atlantic state wanting to become a member."[10]

Kozyrev contended that there was no shortage of simple, superficially appealing solutions to Europe's security problems. "They essentially consist of the extension of Western institutions, whose existence was justified in the Cold War era, to the countries of Eastern Europe. . . . Here the main attention goes to the quickest possible integration of the Central European countries into Western military-political structures."[11]

Such an approach, he stated, was based on two myths: that there was a security vacuum in Central Europe, which would inevitably become the site of either Russian domination or a destructive Russian-German rivalry, and that Russia was moving to adopt a nationalist-imperialist policy in the wake of the December parliamentary elections.

> The solution is produced by taking these two statements together: NATO should protect the countries of Central Europe against the inevitable thrust from the east. The "guard house on the Rhine" should thus be shifted to the banks of the Bug. This request is substantiated by references to the West's "moral obligation" toward the democracies in the region and by reflections about the "natural geopolitical interests" of the Western nations themselves.[12]

Kozyrev criticized that scenario as alarmist and cautioned that if NATO's leaders acted upon such assumptions they would strengthen extremists in Russia who argue for a revival of the Russian empire to resist an attack from the West. A Western strategy based on a worst-case scenario, therefore, was particularly dangerous because it could become a self-fulfilling prophecy.

Such concerns expressed by a moderate like Kozyrev must be taken seriously. If even the most pro-Western elements of Russia's political establishment conclude that they cannot abide an expansion of NATO that excludes their country, one can readily imagine the level of hostility among more stridently nationalist elements.[13] Atlanticists may view NATO's enlargement as benign and stabilizing, but the Russians apparently do not.

Yeltsin and his foreign policy team seem uncomfortable with the entire emphasis on NATO as the institution for dealing with Europe's post–Cold War security problems.[14] Writing in *Foreign Affairs*, Kozyrev gives the Partnership for Peace at best a backhanded endorsement, saying that it answers the need to define Russia's relationship with the alliance "for now." His coolness toward the alliance is evident when he cautions that the program must not "stimulate NATO-centrism among the alliance's policymakers or NATO-mania among impatient candidates for membership." Other comments by Kozyrev suggest a suspicion that proposals for expansion are motivated by animus toward Russia. He charges that both groups seem intent on "ferreting for proof that the Russian government is allegedly changing its foreign policy to suit its nationalist opposition."[15]

Kozyrev and other Russian officials have repeatedly advocated strengthening the Conference on Security and Cooperation in Europe (CSCE) as an alternative to an expanded NATO. According to Kozyrev:

> The creation of a unified, non-bloc Europe can best be pursued
> by upgrading the Conference on Security and Cooperation in
> Europe into a broader and more universal organization. After
> all, it was the democratic principles of the 56-member CSCE
> that won the Cold War—not the NATO military machine. The
> CSCE should have the central role in transforming the post-
> confrontational system of Euro-Atlantic cooperation into a
> truly stable, democratic regime.[16]

His reasons for preferring the CSCE to NATO are not hard to fathom. NATO is an organization dominated by the Western powers, especially the United States, and it has a pronounced military emphasis. Conversely, the CSCE has virtually no military component beyond a vague commitment to the concept of peacekeeping. Its principal focus is on political dialogue and conflict resolution by peaceful means. There is no conceivable way that the CSCE could pose a military threat to Russian interests. Furthermore, because all decisions in the CSCE must be made by consensus, Moscow would exercise a legal veto over policies it disliked. Finally, given its growing network of clients among the other former Soviet republics, Russia could expect to exert far more influence in CSCE councils than it could ever hope to do in NATO.

### The Naivete of NATO Expansionists

Would-be architects of a larger NATO sometimes exhibit an actual or feigned naivete about Russia's probable reaction. Asmus, Kugler, and

Larrabee, for example, chafe at the West's reluctance to involve itself in Eastern Europe's security because of Russia's "strategic sensibilities," adding that "it is hard to understand how supporting democracy and stability in Eastern Europe can undercut democracy in Russia." They further make the extraordinary comment, "Whether NATO's eastward extension becomes a new offer for partnership [with Russia] or a move toward an anti-Russian alliance rests almost entirely on the outcome of Russia's own internal transformation."[17] The Hudson Institute's Gary Geipel states that instead of undermining Russian democrats and playing into the hands of the ultranationalists, "if NATO successfully demonstrates resolve in Central Europe while promoting a more equitable partnership with Moscow, it will strengthen Russian leaders who believe in good relations with the far abroad and weaken those who seek imperial reassertion in the near abroad."[18]

Janusz Onyszkiewicz, a member of the Polish parliament and a former defense minister, states confidently that arguments that the expansion of NATO would isolate Russia and create animosity toward the West can "be dismissed out of hand."[19] Edward L. Rowny, special adviser to presidents Reagan and Bush on arms control issues, exudes a similar degree of optimism. After talking to (unidentified) Russian leaders, Rowny stated: "I am convinced that bringing Central Europe into NATO will not be seen as a military threat. Thoughtful and responsible leaders realize that NATO was—and will continue to be—a purely defensive alliance that threatens no one." He added, however, that "Western leaders need to reiterate this point" so that Yeltsin's opponents "are prevented from exploiting the situation with their paranoid and populist arguments."[20]

Even Zbigniew Brzezinski, who is more sensitive to the probable Russian response than are most advocates of limited expansion, believes that committed Russian democrats have no reason to regard NATO's movement eastward as a hostile act. Indeed, they ought to view that action as being in their own long-term best interest.

> That the expansion of the zone of democratic Europe's security would bring the West closer to Russia is no cause for apology. An eventually democratic Russia should wish to link itself with a stable and secure Europe. Only then will modernity and prosperity become Russia's reality. On this issue, propitiating Russian imperialists is not the way to help Russian democrats. The right course is to insist firmly that the gradual expansion of NATO eastward is not a matter of "drawing a new line" . . .

> but of avoiding a security vacuum between Russia and NATO
> that can only tempt those in Russia who are more than ready
> to opt for empire over democracy.[21]

The import of such statements is that no bona fide democratic government in Russia could possibly regard NATO's enlargement as a hostile act. All reasonable Russian democrats would understand that NATO's motives were benign and, therefore, would not object to having a U.S.-dominated military alliance on the frontiers of their country. Only nasty, authoritarian Russians would harbor suspicions about NATO's intentions, and if such Russians came to power, NATO's leaders would then be justified in transforming the alliance into an instrument of a new containment strategy.

Such attitudes are a manifestation of foreign policy ethnocentrism as well as wishful thinking. They assume that if the United States and the other Western powers regard an expansion of NATO as a benign act to promote stability and democracy in Eastern Europe, Russia has an obligation to view matters the same way. Any deviation from that "proper" perspective is evidence of faulty thinking or, worse, evil intent.

The argument that Russians ought to regard NATO's growth passively also simplistically divides Russians into two stark political camps: peace-loving democrats and aggressive imperialists. That leaves little room for the possibility that some Russians might be (more or less committed) democrats but also strong nationalists who believe that their country should rightfully have zones of influence in regions along its border. The assumption that Russia has the right to act as a normal great power in the international system is not necessarily a manifestation of unbridled imperialism. Yet that appears to be the operating premise of many NATO partisans.

Responding to the Asmus, Kugler, and Larrabee article, Alexei Pushkov, deputy editor of *Moscow News*, excoriated their reasoning. "Expanding NATO membership to include East European countries while leaving Russia aside," he warned, "would be playing with fire." Pushkov ridiculed the notion that enlargement proposals were motivated only by the desire to assist fragile democracies and market economies in the region and to promote general stability. "Military-political alliances do not exist for the sake of abstraction or charitable purposes. They are always directed against someone or something." In this case, the inescapable conclusion "is that NATO expansion to

Eastern Europe can be directed only against one country: Russia." He dismissed assurances to the contrary. "The security and economic 'confidence-building' measures that the RAND analysts offer up to allay Russian concerns are designed to justify the ugly-looking idea of a new *cordon sanitaire* between the West and Russia."[22]

Andranik Migranyan, a member of Yeltsin's presidential council, expressed similar suspicions about Washington's motives for wanting to extend NATO eastward.

> Clearly Russia has reasons to oppose NATO's expansion to its borders. The U.S., through the alliance, intends to preserve its military and political leadership in Europe. The expansion of NATO—initially through the Partnership for Peace—is a real step on the way to filling the power vacuum [in Eastern Europe], with the ultimate goal of restraining and disciplining Russia itself.[23]

Moscow would, in fact, have good reason to worry about NATO's enlargement. Great powers have usually been more concerned about competitors' capabilities than about their intentions—because intentions can change quickly. And for all the talk of NATO's being largely a political organization in the post–Cold War era, it remains a military alliance with impressive capabilities. Even Russians who are not closet aggressors might be uneasy about having such a powerful military association perched on Russia's borders. They would have to wonder, for example, what NATO's response might be if Moscow continued to oppose aspects of Western policy toward the former Yugoslavia, where Russia has long-standing ethnic and religious ties as well as other interests. A large NATO bloc would, at the very least, be an intimidating presence, guaranteeing that Russia would have little choice but to go along with the alliance's policy preferences—even if Russian leaders believed the policies to be misguided, unjust, or contrary to Russia's own best interests.

Even the most peaceably inclined great power would chafe at such constraints and the humiliating subordination they symbolize. To place the matter in perspective, imagine what the U.S. response would be if a Russian-led Warsaw Pact, heretofore confined to Eastern Europe, decided to exploit a temporarily weakened American geostrategic position by incorporating Western Europe and Canada into an enlarged alliance. Even if Russia and all its allies had impeccable democratic capitalist credentials and offered repeated assurances of the most peaceful, friendly motives, it would be difficult to imagine the United States' interpreting the action as anything other than an encirclement strategy. We should not expect Russia to view an expanded NATO any differently.

Given Russia's current weakened condition, it might be possible to force Moscow to make such a humiliating capitulation. But the long-term cost, in terms of both the impact on domestic political developments in Russia and the destructive effect on the West's relations with that country, might prove much higher than any reasonable official would want to pay. A NATO-designed *cordon sanitaire* would undermine the relatively pro-Western Yeltsin government and confirm the allegations of Zhirinovsky and the other ultranationalists who already denounce Yeltsin and Kozyrev as Western stooges. Pushkov warns that "attempts to isolate Russia, to throw it back beyond Europe's confines, in the end only play into the hands of those anti-democratic forces in Russia who tend to view the West with a lot of distrust or even animosity."[24]

### The Risky Strategy of Trapping the Russian Bear

Eager NATO expansionists would be wise to heed Pushkov's warning. Although widespread anger about the dismal state of the Russian economy was the single most important reason for the unexpectedly strong showing of Zhirinovsky's party in the December 1993 parliamentary elections, a less prominent but still significant factor was the deep sense of national frustration with and humiliation about Russia's position in the world. That factor continues to have explosive potential.

In retrospect, it is evident that the Soviet Union's superpower status was largely a fraud. At most, the USSR was a one-dimensional superpower. Its claim to global influence was based almost entirely on its perceived military capabilities. Economically, the country was far weaker than even the most skeptical Western observers believed at the time. Contrary to private estimates and CIA assessments that portrayed the Soviet economy as approaching the $3-trillion-a-year mark by the late 1980s, the evidence now indicates that at no time was it more than half that size.

During the USSR's twilight years, doubts surfaced even about the country's conventional military capabilities. The Red Army's surprising inability to subdue Afghanistan indicated that the Soviet military may have been something less than the modern-day equivalent of Nazi Germany's vaunted *Wehrmacht*. Increasingly, it appeared that Soviet power derived from one thing alone: Moscow's arsenal of 30,000 to 45,000 tactical and strategic nuclear weapons. The caustic comment that a few Western analysts made about the Soviet Union throughout the Cold War—that the USSR was merely a Third World country with nukes—was apparently very close to the mark.

Although the USSR's superpower credentials may have been bogus, most Russians believed they were genuine. It has been an extraordinarily traumatic experience for them to watch their once-proud country—the feared rival of the West—become an international beggar, dependent on Western financial largesse. Almost as galling is the realization that while Russia may be treated with indulgence in important international gatherings, it is seldom treated with much respect.[25] The Yeltsin government's inability to gain membership in the Group of Seven industrial powers and the pattern of forced deference to U.S. policy views on a host of issues, from Iraq to Yugoslavia, that impinge on important Russian interests have been especially potent sources of humiliation. Zhirinovsky and other nationalists have skillfully exploited public anger at their country's precipitous decline in status and at a regime that seems unable or unwilling to "stand tall" for Russia.

Those in the West who fear that Zhirinovsky has the potential to become another Hitler ought to reflect on the factors that helped bring the Nazis to power in Germany. Hitler was able to take advantage not only of the dreadful economic conditions that afflicted Germany during the Great Depression—probably the decisive factor—but also of the pervasive atmosphere of resentment that Germany was treated as a second-class power in the international community. Anger at the inequities of the Treaty of Versailles—especially the "war-guilt" clause that blamed Germany alone for the tragedy of World War I and the reparations requirements, which led to the hyperinflation of the mid-1920s and the financial devastation of the German middle class— became political kindling that helped produce the Nazi conflagration. Intransigence on the part of the victorious allies regarding even the most basic and legitimate German objectives (such as an *Anschluss* with Austria, something that was desired by clear majorities in both countries during the years immediately following World War I) also produced festering grievances.

The allies' myopic policy of alternately treating democratic Weimar Germany as an international nonentity and as a probable future aggressor led to tragedy. Those who are tempted to take advantage of Russia's weakened condition risk causing a similar tragedy. Disregarding Moscow's long-standing interests in Eastern Europe and enlarging NATO for the purpose of keeping Russia passive and subordinate create ideal conditions for a backlash. If the West wants to help bring Zhirinovsky or one of his ilk to power—or more likely a less bizarre and flamboyant neo-imperialist such as former vice president Alexander Rutskoi—it could scarcely adopt a better strategy than expanding NATO to Russia's borders.

### Evaluating the Russian "Threat"

Concerns about a possible resurgence of Russian power and influence are not entirely misplaced, but they also should not be exaggerated. Moscow has already taken a number of steps to increase its political, economic, and even military penetration of some of the neighboring republics. The Yeltsin government's effort to strengthen the Commonwealth of Independent States (CIS), the establishment of an expanding ruble zone in which the Russian currency is the dominant medium of exchange, and the dispatch of "peacekeeping" forces to such trouble spots as Moldova, Tajikistan, and Georgia are all manifestations of that trend.[26]

The strategy has clearly met with some success. Several former Soviet republics are now included in the ruble zone, the incumbent governments in Tajikistan and Georgia are heavily dependent on Moscow's military forces for protection from powerful insurgent movements, and the CIS has endorsed a vaguely defined peacekeeping role for Russia in the "near abroad" as well as a variety of initiatives for military cooperation. There is little doubt that such actions limit the independence of some of Russia's neighbors. Other indications of the Kremlin's intent include a request to the United Nations to give a blanket endorsement of Moscow's authority to conduct peacekeeping operations in the near abroad and the proposal (since withdrawn) that the Conventional Forces in Europe Treaty be changed to authorize the deployment of additional Russian military units along the country's southern frontier.[27]

Those are all signs of a great power attempting, despite its current distress, to flex its political muscles on matters considered important to its national security. Nevertheless, there is little cause for panic. Moscow's ability to establish and sustain the ruble zone is testimony more to the extreme economic weakness of the other republics than to any Russian economic strength. Russia's economy, in fact, continues to be plagued by inflation, declining production, and an assortment of other woes. It is unlikely to be an effective vehicle for influence in the rest of Eastern and Central Europe for a very long time. Indeed, Russia needs financial help from other members of the CIS to fund peace-keeping operations in the near abroad and has complained bitterly that more assistance has not been forthcoming.[28] That is scarcely the action of a confident, would-be regional hegemonic power, nor are the unresponsive policies adopted by the smaller CIS countries consistent with the deference that would be shown by puppet states.

Moscow's military strength is not much greater than its economic strength. Russia's conventional military forces are in severe disarray and its military budget continues to plunge; spending may fall as low as $20 billion for the next fiscal year, and the number of active duty military personnel is slated to be cut to 1.9 million.[29] The dramatic slashing of the defense budget that has taken place in recent years is inconsistent with an ambitious neo-imperial strategy. Another indicator of the Kremlin's limited objectives is that instead of trying to maintain huge tank armies and other power projection capabilities, the military's new strategic plan features lightly armed, rapidly deployable airborne troops and peacekeeping units.[30] Those forces might be useful for small-scale missions in the near abroad, but they would not be effective for any major expansionist push into the former Warsaw Pact countries, much less into Western Europe.

Barring the emergence of an ultranationalist (or less likely a neo-Leninist) government, Russia can be expected to act like a typical great power as it slowly recovers its political and economic strength. A conventional great power is not likely to embark on an unlimited expansionist binge, although it may eventually strive to establish a zone of influence. Both the limited and the broad anti-Russian versions of NATO expansion are predicated on the belief that the West cannot tolerate such a development.[31] But is that a valid assumption?

## Tolerating a Russian Sphere of Influence

It would, of course, be desirable in the abstract if great powers never pressured or intimidated their smaller neighbors. Indeed, it would be ideal if all states in the international system had democratic governments with strictly limited powers and displayed an abiding respect for the rights of all individuals living in their jurisdictions. Under such conditions, national boundaries would cease to have much significance, since commerce would flow freely and the citizens of one country would enjoy essentially the same rights as citizens of all other countries. "Nations" would then be little more than expressions of benign cultural differences.

Unfortunately, that idealistic vision bears little resemblance to the actual characteristics of the international system at any point in time. It certainly does not correspond to reality now or in the foreseeable future. We must, therefore, consider less than ideal policy options. Because great powers—including the United States, as most Latin Americans will readily attest—typically do carve out spheres of influence and coerce smaller neighbors that resist hegemonic dictation, the pertinent question is whether—or to what extent—the United States and the nations of Western Europe can tolerate such behavior on the part of Russia in Eastern Europe. Given the greater proximity of the West European powers to Moscow's prospective zone of influence, the answer may be significantly different for them than it is for the United States.

From the U.S. perspective, a Russian zone of influence in Eastern Europe should not necessarily pose a security threat. Russia, in fact, did exercise a considerable amount of influence—in competition with Austria-Hungary—in much of that region during the 19th and early 20th centuries without having the slightest discernible adverse impact on American interests. Even Moscow's heavy-handed domination of Eastern Europe during the Cold War did not per se threaten important American interests. The reason for U.S. alarm was that the Soviet Union might use its control of Eastern Europe as a springboard for establishing a hegemonic position in Western Europe.

It is imperative that U.S. policymakers make the distinction between an aggressively expansionist state with seemingly unlimited ambitions subjugating Eastern Europe as an apparent prelude to a bid to control one of the major global military and technological power centers and a normal great power's goal of having a zone of influence in its immediate neighborhood. The former situation—which is an approximation of Moscow's conduct after World War II—would be cause for concern; the latter can be accepted without great trepidation.[32]

There is no inherent reason a decision by a nontotalitarian Russia to implement its version of a Monroe Doctrine in portions of Eastern Europe and other areas along its border would menace America. Such action would no more pose a threat to U.S. security than Washington's hegemonic position in Latin America and the Caribbean—including the use of military coercion against nations in those regions on more than a few occasions—has threatened Russia's security. Indeed, to the extent that Washington is concerned about the possibility that instability in Eastern Europe could cascade westward, a Russian regional policing role might marginally serve American interests by promoting greater stability. Moscow's involvement has already at least arguably dampened conflicts in Georgia, the transdniester region of Moldova, and elsewhere in the near abroad.

Preventing the establishment of a Russian sphere of influence certainly should not be so important to the United States that Americans would be willing to risk a major war by extending NATO's security guarantees to the nations of Central and Eastern Europe. Max Kampelman's assertion that no great power has "a moral or political right to determine the foreign policy, much less the internal affairs, of Eastern European democracies" is a noble invocation of idealistic values, but it is also profoundly out of touch with long-standing realities of international politics. Merely asserting, as he does, that the international community must "make clear that the day of hegemonic spheres of influence is over" begs the question of how such a policy can be enforced at a tolerable level of risk.[33]

Both geographic and economic factors more severely constrain the ability of the West European states to tolerate a Russian zone of influence in Eastern Europe. Russia's ambitions in that region, even if they remain relatively restrained, cannot be a matter of complete indifference to the members of the European Union, especially to the EU's easternmost major member, Germany. At some point, Moscow's definition of its sphere of influence may begin to intrude on what the

EU powers regard as their zone of safety. There is some possibility of friction as the competing powers attempt to resolve or at least manage that problem.

Several outcomes are theoretically possible. Given Russia's current weakness, the EU may be able to establish such an extensive web of political and economic ties throughout Eastern Europe that it will be difficult for Moscow to subsequently gain much of a foothold without being willing to risk a military confrontation. Conversely, if Russia's recovery is relatively rapid and it adopts an assertive foreign policy early on, it may be able to compete on equal, or perhaps more than equal, terms. That would mean either a contentious rivalry or an agreement (explicit or implicit) between Russia and the EU to divide the region into separate spheres of influence. Finally, it is conceivable that the Central and East European states may unite sufficiently to minimize domination by either the EU or Russia, although given the weakness of those countries and their history of divisiveness, that is the least likely scenario.

The different U.S. and EU thresholds of tolerance for Russia's sphere of influence underscore why NATO is an inappropriate institution for dealing with that issue. If the alliance adopted as one of its objectives the exclusion of Russian power from Eastern Europe, the West European states would be inclined to advocate a hard-line response even when there was no danger to America's security. That is only one of many instances in which the notion of a united "transatlantic perspective" is illusory, but it is one with an exceptionally high potential for danger. An effort by NATO to implement a "containment-plus" strategy against Moscow would involve a substantial danger of war. If the West Europeans deem it in their interests to incur such a risk, they should do so through their own security alliance. The United States would reap little benefit by joining them in such a perilous venture.

### Sources of Tension

Even in the absence of a concerted effort by Moscow to rebuild the Soviet empire—or at least carve out a Russian sphere of influence in the near abroad—there are numerous unresolved problems that could lead to limited Russian military action against neighboring states. Those possibilities make a NATO security commitment to Russia's neighbors extremely hazardous.

Relations between Russia and the newly independent Baltic republics, for example, are tense. Russian nationalists are not reconciled to the loss of territory that had been part of the empire since the time of Peter the Great. Ill-advised actions by some of the Baltic governments have exacerbated tensions. Viewing Russian residents as symbols of Soviet imperialism, Estonia and Latvia have created severe legal hurdles that ethnic Russians must clear before they can enjoy the rights of full citizenship.[34] Moscow has reacted angrily to such discrimination, at one point even threatening to delay the withdrawal of its military units from Estonia, and has asserted the right to intervene, if necessary, to protect the rights of Russians. Although Latvia modified its harsh citizenship law in July 1994, responding to intense pressure from Moscow, there are still features of that statute as well as other measures that flagrantly discriminate against Russian residents.[35]

The underlying hostility toward Russians in all three Baltic states has not noticeably diminished. It was indicative that when Clinton, in his July speech in Riga, urged Latvians "never to deny to others the justice and equality you fought so hard for and earned for yourselves," he was greeted with a stony silence.[36] Tensions are likely to persist, and it is not far-fetched to envision a scenario in which Russia resorts to force in response to actual or perceived mistreatment of its ethnic cousins living in the Baltics. Nationalistic politicians in Russia are already demanding that the Yeltsin government take stronger action. Ethnic Russians in Estonia are treated as "10th-class people, not even second class," fumed Vladimir F. Shumeiko, president of the upper house of the Russian parliament during a visit to Washington, adding that "we cannot sacrifice the Russians who remain in Estonia."[37] Even Andrei Kozyrev states categorically that "protecting the rights of the Russian-speaking population [in those republics] shall continue to be a high priority objective."[38]

The most ominous situation, though, is the confrontation between Russia and Ukraine. There are numerous sources of friction, and disputes have already flared over such issues as the control of the Black Sea fleet and the disposition of Soviet nuclear weapons remaining on Ukrainian territory. Ukraine's parliamentary elections in the spring of 1994 also revealed a disturbing pattern of ethnic division. Voters in the "Russified" eastern portion of the country voted heavily for parties, including one dominated by former communists, that favored closer ties with Moscow. The western sections of the country elected candidates of Rukh and other strongly Ukrainian nationalist parties. Even before that election, the CIA reportedly warned the Clinton administration that there was a serious possibility of Ukraine's fragmentation, perhaps accompanied by civil war.[39]

The most troublesome and dangerous issue of all is the status of the Crimea—which is largely inhabited by Russians but is now under Kiev's control.[40] The election of an outspoken Russian nationalist, Yuri Meshkov, as the head of the Crimean provincial government in January 1994 noticeably increased the potential for trouble. Tensions were ratcheted up several notches further two months later when Meshkov's administration conducted an unauthorized referendum—officially dubbed an "opinion poll" to circumvent the Ukrainian parliament's prohibition—that recorded an overwhelming vote in favor of secession.[41] The election of Leonid Kuchma as Ukraine's new president in July 1994 may dampen those tensions somewhat, since Kuchma received a majority of his support from the Russified eastern region of the country. Nevertheless, the roots of the problem are deep and Ukraine's unity remains fragile at best.

Ukrainian nationalists are nervous not only about the secessionist rumblings in the Crimea (and in the eastern provinces as well), they also worry about Moscow's intentions. Their apprehension is well-founded. Russians of all political persuasions seem to adopt irredentist attitudes when it comes to the Crimea. Even St. Petersburg mayor Anatoly Sobchak, a staunch pro-Western democrat, contends that the Crimea is rightfully Russian soil, and he believes that Russia should claim the borders it had in 1922, when it joined the Soviet Union.[42] Such territorial claims are symptomatic of a larger problem: Russians across the political spectrum have trouble accepting the legitimacy of an independent Ukraine. To them, Ukraine is the heart of Rus, the original medieval Russian kingdom that was conquered by the Mongols in the 12th century and not fully restored to Russian control until the 18th century. Of all the losses of territory occasioned by the breakup of the Soviet Union, that of the Ukraine has been the most emotionally wrenching for Russians.

Most Ukrainians are understandably suspicious of Russian intentions, and they are determined to preserve their country's hard-won independence. One obvious manifestation of Ukrainian apprehension has been Kiev's reluctance to relinquish the nuclear weapons that it inherited from the Soviet Union. The price Ukrainian officials demanded for giving up that arsenal was not only substantial Western economic aid, but also reliable security guarantees.[43] They have made no secret that they want NATO's protection and, if possible, NATO membership.

The tripartite agreement signed by Clinton, Yeltsin, and then Ukrainian president Leonid Kravchuk in Moscow in January 1994 under which Kiev will eliminate its arsenal in stages over seven years does not necessarily end the nuclear weapons dispute. Aside from the fact that the Ukrainian parliament may ultimately repudiate or undermine Kravchuk's handiwork, seven years is a very long time; a number of developments could scuttle final implementation of the accord. Kuchma's election significantly increases that probability. As prime minister in 1993, he repeatedly sided with parliamentary forces who wanted to retain the nuclear weapons. Within days of his election victory in July 1994, he suggested that Ukraine reconsider its declared intention to join the Nuclear Nonproliferation Treaty (NPT) as a nonnuclear state and indicated further that Kiev would insist on more sizable aid flows from Russia and the West before it would relinquish the nuclear warheads.[44]

Given that political context, the security assurances to Kiev contained in the accord ought to concern Americans. Clinton and his principal advisers insisted that the United States and Russia only provided the usual assurances given to signatories of the NPT and the CSCE. The text of the trilateral statement by the three presidents issued at the signing ceremony does little to clarify the matter. The three pertinent clauses stated that the United States and Russia will

> Reaffirm their commitment to Ukraine, in accordance with the principles of the CSCE Final Act, to respect the independence and sovereignty and the existing borders of the CSCE member states and recognize that border changes can be made only by peaceful and consensual means; and reaffirm their obligation to refrain from the threat or use of force against the territorial integrity or political independence of any state, and that none of their weapons will ever be used except in self-defense or otherwise in accordance with the Charter of the United Nations;
>
> Reaffirm their commitment to seek immediate UN Security Council action to provide assistance to Ukraine, as a non-nuclear-weapon state party to the NPT, if Ukraine should

become a victim of an act of aggression or an object of a threat
of aggression in which nuclear weapons are used;

Reaffirm, in the case of Ukraine, their commitment not to use
nuclear weapons against any non-nuclear-weapon state party
to the NPT, except in the case of an attack on themselves, their
territories or dependent territories, their armed forces, or their
allies, by such a state in association or alliance with a nuclear
weapon state.[45]

On their face, such assurances would appear to be little more than
bland paper promises. But there are indications that the security
assurances contained in the Ukrainian agreement may go beyond
standard disarmament boilerplate.

Precisely what the tripartite agreement means in terms of substantive U.S. obligations is not clear. William Miller, the U.S. ambassador to
Ukraine, offered his view that the commitment meant only, "If any
state carries out an attack on Ukraine or makes territorial claims on it,
the United States will appeal to international organizations such as the
United Nations in order to protect Ukraine."[46] Administration leaders
also continue to insist that there is no possibility the United States
would use force to defend Ukraine's sovereignty or territory.

But Ukrainian officials apparently believe that they have received
meaningful promises from Washington, stating privately that their
country deemed it "critical" that security assurances be contained in a
document signed by the United States as well as Russia.[47] Moreover,
Kiev had insisted repeatedly before the January accord that the
standard assurances given to NPT and CSCE signatories were not
sufficient. Granted, Ukraine's dire economic straits—especially the
desire for Western financial aid and cheap energy supplies from
Russia—created powerful incentives for Kiev to accede to Washington's wishes on the nuclear weapons issue, but it seems unlikely that
Ukrainian officials capitulated on that crucial point. In particular, one
wonders why they would be content with a security "guarantee" that
involved nothing more than a U.S. appeal to the United Nations—
where Russia has a veto in the Security Council.

It is not reassuring that U.S. officials admit that the treaty contains
several secret provisions.[48] The arrogance of administration leaders on
the issue of secrecy is breathtaking. One senior official told the news
media that the United States would not disclose what "incentives" had
been offered to Ukraine beyond those vague provisions already made
public. "It is our position that some elements of the agreement will

remain confidential," the official confirmed, adding that there was no timetable for making the secret clauses public and that they might, in fact, never be publicly released.[49]

Americans have every right to know what security assurances the Clinton administration has given and whether such promises create even the remote risk of U.S. involvement in a Russian-Ukrainian conflict. Syndicated columnist Mona Charen, citing the reports that the United States had offered assurances that Ukraine's borders would be permanent, concluded, "That sounds like it might mean a U.S. commitment to defend Ukraine if she is attacked."[50] Charen is probably overstating matters, but given the aura of secrecy, there is no way to be certain.

If the United States and its allies offered NATO protection, or even informal security assurances, to induce Kiev to relinquish its nuclear arsenal, the alliance could become entangled in a regional imbroglio that might explode into an armed struggle at any time. NATO would then be confronting a nuclear-armed Russia over stakes far less substantial than those for which it assumed similar risks during the Cold War.

Even Charles L. Glaser, a cautious advocate of NATO expansion, concedes that "tensions and disagreements between Russia and Ukraine highlight the possibility of a major power war in the East." Moreover, "because war in the East is likely, Western commitments would likely be put to the test."[51] Yet most proponents of new NATO missions in Eastern Europe act as though there is no prospect that such security promises will ever have to be redeemed.

It is on that issue that NATO expansionists of all types tend to be the most evasive. They insist that alliance security commitments would prevent a repetition of Russian expansionism and thereby enhance the stability of the region. Yet (with the exception of Glaser and a few others) they also seek to minimize both the likelihood and the severity of the risks the United States and its alliance partners would be incurring if the alliance moved eastward. Such a position is inconsistent if not disingenuous. Either the alliance intends to afford the nations of Eastern Europe reliable protection against Russian expansion or it does not.[52] If the former is true, the commitment involves very serious risks. If the latter is the case, NATO's leaders are engaging in an appalling act of deceit that could prove fatal to any East European nation foolish enough to rely on the alliance.

That is an issue that simply cannot be finessed. William E. Odom, former director of the National Security Agency, argues that NATO could constrain Russia's "imperial reconsolidation" if the alliance moved its boundary to the eastern borders of Poland, Slovakia, and Hungary. "It would also reassure the Baltic states by bringing NATO forces closer to them. And it might even affect the capacity of Ukraine to sustain its independence." Having NATO "close at hand," Odom insists, "would affect the political psychology in the belt of states between the Baltic and the Black Seas."[53] That is an insidiously dangerous case of wishful thinking. Odom's analysis assumes that the United States and the other members of NATO can issue promissory notes of security to the nations of Central and Eastern Europe, largely for the purpose of psychological reassurance, without having to worry that those notes will ever be presented for payment. But to be effective, a military alliance must be a credible security shield, not a psychological security blanket. Expansion based on the latter assumption is little more than an irresponsible bluff that Russia, given its extensive interests in Eastern Europe, might someday decide to call.

The ongoing Russian-Ukrainian imbroglio illustrates the potential dangers entailed in expanding the North Atlantic alliance into Eastern Europe. *National Interest* editor Owen Harries, a critic of enlargement, sharply questions the logic of expansion.

> The proposal takes no account at all of Russian susceptibilities and interests and envisages no role for Russia in Eastern Europe. NATO is simply to take over responsibility for the stability of a region that has been in Russia's sphere of influence for centuries. The 45-year interlude of the Soviet bloc was merely an episode in a much larger history, and its demise should not be taken as marking the end of Moscow's involvement. Strategic interests, traditional motives of prestige, the "historic mission" of freeing the Greek Orthodox population from infidel rule, and the pan-Slavism that had a varying but real impact on policy—all these combined to make Eastern Europe . . . a matter of intense concern for Russia long before Lenin and Stalin appeared on the scene.[54]

Harries warns that to "ignore all this history and to attempt to incorporate Eastern Europe into NATO's sphere of influence, at a time when Russia is in dangerous turmoil and when that nation's prestige and self-confidence are badly damaged, would surely be an act of outstanding folly."[55]

65

## A Bogus Solution: Comprehensive Expansion

Some advocates of enlargement seem to realize, at least implicitly, that Russia might well be suspicious of NATO's incorporation of the East European countries. American University political scientists William Kincaide and Natalie Melnyczuk caution that the form and extent of the alliance's expansion could be crucial. Even the Partnership for Peace risks "creating a Europe not from the Atlantic to the Urals but to an extended Curzon Line—the border between Poland and Russia drawn by the Western Powers in the 1919 Versailles Peace Treaty without Russian participation. Such a division of the region would create a new and troublesome fault line."[56] James Baker likewise acknowledges that warnings that the alliance's eastward thrust could antagonize Russia "would have credence if NATO expansion were to include Central and Eastern European states but exclude the states of the former Soviet Union."[57]

Some policymakers who are concerned about a possible hostile reaction from Moscow have sought to avoid the problem by simultaneously strengthening the West's security links to Russia. Leslie H. Gelb, president of the Council on Foreign Relations, attempts to be both hard-line and conciliatory toward the Russians. "If stability and democracy in Eastern Europe require a NATO security shadow, it should be lengthened with or without Moscow's blessing. To be sure, Russian leaders should not be made to feel left out of European security arrangements. Nor, more importantly, should they be left with the idea that they can intimidate their European neighbors once again."[58] How the great tension between such competing objectives could be resolved, Gelb does not say.

Zbigniew Brzezinski offers a more substantive scheme to overcome the dilemma, suggesting "a far-reaching NATO proposal for a treaty of alliance with Russia" to accompany an expansion of NATO's membership.[59] Brzezinski strives even harder than Gelb to have it both ways.

> The deliberate promotion of a larger and more secure Europe need not be viewed as an anti-Russian policy, for the inclusion in NATO of several Central European democracies could be coupled with a simultaneous treaty of alliance and cooperation between NATO and Russia. It is altogether unlikely that Russia could be assimilated into NATO as a mere member without diluting that alliance's special cohesion—and that is

certainly not in America's interest. But a treaty between the alliance and Russia (even if Russia falls short of U.S. hopes for its democratic evolution) would provide the Russians with a gratifying recognition of their country's status as a major power while embracing Russia within a wider framework of Eurasian security.[60]

Baker and other comprehensive expansionists pursue an even more grandiose goal: a new NATO that would ultimately include Russia as a member. Jeane Kirkpatrick contends, "All the reasons that NATO membership—or its equivalent—should apply to Eastern Europe apply as well to Russia."[61] *The New York Times* agrees, arguing that "if NATO opens its doors to new members, Russia should be in line along with Poland, the Czech Republic and Hungary. The prospect of NATO membership could encourage the Russian Army to become a force for peace; that alone would justify reshaping an anti-Russian alliance into an altogether different security structure."[62] Max Kampelman advocates the prompt admission of the Visegrad countries, but contends that with "time and experience" the enlargement could encompass other states in Eastern Europe, "including any of the ex-Soviet states, such as Russia."[63]

Including Russia in NATO's embrace is not a new idea. Australian political scientist Coral Bell advanced a similar proposal nearly four years ago, before the collapse of the Soviet Union.[64]

Such schemes are even fuzzier and more ill-conceived than the usual models of an enlarged NATO. Bringing Russia into the alliance would create a cumbersome political-military association spanning three continents. Such an unwieldy organization would make a farce of the concept of a coherent military alliance. Henry Kissinger asks the pertinent question, "If everybody between Vladivostok and Vancouver is eligible—if everybody is allied with everybody—why even have an Atlantic Alliance?"[65] Former secretary of state Alexander Haig, like Kissinger a proponent of expansion without Russia, is more blunt: "To include Russia in NATO is a geostrategic oxymoron."[66]

Indeed, that incarnation of NATO would have a membership virtually congruent with the CSCE. If the goal is the establishment of such a vast pan-European (or more accurately Euro-Asian) security organization, it would seem to make more sense to strengthen the powers of the CSCE than to have NATO duplicate the conference's territorial coverage. Realistically, it is difficult to envision such a cumbersome entity as a Euro-Asian NATO playing an effective security role in any case. On an operational level, the difficulties of incorporating Russia's military into NATO's defense arrangements would make the formidable problems of integrating the militaries of the East European states pale in comparison.

But there is a more important reason why including Russia in NATO would be the pinnacle of folly. Such a step would inject the alliance into a multitude of conflicts already raging in the Caucasus, Central Asia, and other places on the perimeter of the former Soviet Union. It would be bad enough to involve NATO in the ethnic problems of Eastern Europe, but it would be far worse to entangle the alliance in the wars taking place in Georgia, Azerbaijan, or Tajikistan.[67] Yet Russia as a NATO member would have every right to call on its allies for assistance in dealing with such problems.

Making Russia a signatory to the North Atlantic Treaty could entail even more serious dangers. There are long-standing territorial disputes between Russia and China, and although tensions have eased somewhat in recent years, large numbers of troops are still deployed on both sides of the border. Indeed, one ought to ponder how China might react to a NATO-Russian military presence on its doorstep. Kissinger cautions that the Partnership for Peace—to say nothing of NATO with Russia as a formal member—could be viewed in Asia as "ranging the white races against the peoples of the Far East."[68]

That prospect again raises questions about the credibility of NATO's security guarantees. The more extensive the obligations, particularly when they involve obscure conflicts in remote regions, the less credible NATO's promises of mutual assistance would become. Russian officials such as Migranyan are already citing that problem as one reason why the inclusion of Russia in NATO would be undesirable for all concerned. As a member of NATO, Migranyan points out: "Russia would become the alliance's outpost on its borders with the Islamic world and China. It is hard to imagine that American soldiers would defend this border as they once defended the one dividing East and West Germany."[69]

There is no feasible version of NATO's enlargement that adequately deals with the Russia factor. Both the limited and broad forms of expansion are motivated by animus against Moscow and represent ill-conceived efforts to forestall the reemergence of any Russian sphere of influence in Eastern Europe. They have the potential to be a blueprint for war with Moscow over geostrategic stakes far more modest than those that were at the center of the Cold War.

Comprehensive expansion is a desperate attempt to evade such problems and neutralize Russia's objections to an enlarged NATO. But mere alliance membership could hardly be relied on to dissuade Moscow from taking coercive action against a regional adversary—including a fellow treaty signatory—if Russian leaders decided that important national interests were in jeopardy. Advocates of comprehensive expansion should also ponder the consequences if Russian democracy were to falter after the country had been admitted as a member. That is hardly a remote possibility since Russia's democratic institutions are fragile. How would NATO deal with a treaty ally headed by someone like Rutskoi, who regards the alliance with undisguised hostility?

Comprehensive expansion creates yet another dilemma. The most likely outcome of including Russia would be the erosion of NATO's credibility as it sought to offer security guarantees that neither the nations of Western Europe nor the United States would have any rational reason for wanting to honor. In the improbable event that NATO did seek to honor such far-flung commitments, that effort would ensnare the alliance in an assortment of new controversies and conflicts in Central and East Asia, arousing in the process the suspicions of the countries in those regions and perhaps creating enemies where none currently exist.

# 4. Parochial East European Entanglements

In addition to deterring or co-opting Russia and thereby preventing Moscow from taking coercive measures against neighboring states, another mission for a larger NATO would be to prevent or suppress conflicts among the East European countries themselves. A few proponents of enlargement embrace the first mission but not the second. University of Chicago professor Charles L. Glaser, for example, states that "Western security interests call only for rather narrow security guarantees that are designed to deter Russian expansion into Central Europe, but not to prevent all wars between the smaller countries of Central Europe."[1] Glaser's relative restraint is very much an exception, however, in the expand-NATO camp.

Atlanticists have long argued that one important collateral benefit of NATO during the Cold War was that it suppressed potential conflicts in Western Europe. That unaccustomed atmosphere of security, in turn, enabled the West Europeans to cooperate on an array of political and economic matters and produce a period of unprecedented harmony and prosperity. The alliance—and especially America's unchallenged political and military leadership of it—became, in the words of German foreign policy expert Josef Joffe, "Europe's pacifier."[2]

NATO partisans contend that preservation of the alliance, and the U.S. connection, is essential in the post–Cold War period to prevent the "renationalization" of defense policies throughout Western Europe—which might reignite the deadly national rivalries that led to so many bloody conflicts in the past.[3] Glaser contends that "NATO can play a major role in hedging against the growth of security competition in Western Europe by preserving America's role as a 'defensive balancer.' NATO enables the United States implicitly to promise to protect all countries of Western Europe against one another."[4]

Advocates of expansion often take the pacifier argument one step further. They maintain that if NATO was able to quell intramural disputes during the Cold War and foster an atmosphere of security and cooperation in Western Europe, an enlarged NATO will be able to

71

perform a similar service for the embryonic democracies of Central and Eastern Europe. President Clinton apparently endorses that assumption. In a speech to the Polish parliament on July 7, 1994, the president stressed not only the inevitability of NATO's expansion but that enlargement of the alliance "will not depend upon the appearance of a new threat in Europe. It will be an instrument to advance security and stability for the entire region."[5]

### Can NATO Stabilize Eastern Europe?

There are several problems with that thesis. First, it is not clear that the absence of conflict among the European members of NATO throughout the Cold War was due solely, or even primarily, to the alliance's ability to mediate potential quarrels and eliminate perceptions of insecurity. There may have been a more immediate reason for the lack of intramural conflicts: the realization on the part of most West Europeans that they faced an extremely dangerous common enemy, the Soviet Union, and that fratricidal struggles could prove fatal to them all. It is premature, at the very least, to conclude that alliance solidarity will persist during the post–Cold War era under a wholly different set of conditions.

Second, even if one accepts the argument that NATO has played a crucial role in pacifying Western Europe, there is no guarantee that it would enjoy comparable success in Eastern Europe. Many of the causes of conflict in Western Europe had been ebbing even before the formation of NATO. The Anglo-French antagonism, the complicated dynastic rivalries, and the competition for influence in such inchoate nation-states as Germany and Italy that had once plagued the western portion of the Continent had already become largely historical matters. The principal remaining source of tension—the Franco-German territorial dispute over Alsace-Lorraine—had also been decided emphatically in France's favor after Germany's crushing defeat in World War II. True, French leaders remained apprehensive about a possible resurgence of German power—and NATO undoubtedly helped alleviate that festering problem by harnessing West German military capabilities within a multilateral institution. Nevertheless, the relative absence of virulent disputes in Western Europe after World War II contributed to the ease with which NATO was able to pacify the region.

The alliance would encounter radically different conditions in Eastern Europe. That region is a cauldron of unresolved territorial disputes;

intense ethnic and religious rivalries; and fragile, unstable political systems. The process of nation building in Eastern Europe today resembles the same process in Western Europe three or four centuries ago, with all the attendant brutality and intolerant forms of nationalism. The Soviet Union attempted to act as Eastern Europe's pacifier at the same time NATO played that role in Western Europe—with a notable lack of lasting success. Perhaps an expanded NATO, as an association of democratic states, would fare better, but there is certainly no guarantee.

That uncertainty highlights a final problem with the "alliance as pacifier" thesis. Expansionists assume as a matter of faith that NATO would be able to prevent conflicts and resolve disputes in Eastern Europe as it has done in the West. That assumption begs the question of what the United States and the other current members of the alliance are to do if conflicts break out despite their best efforts at deterrence and pacification. Before they take the fateful step of expansion, the existing signatories must determine whether they are really willing to bear the consequences if their role as pacifier proves to be far more difficult and costly than anticipated.

## Distinguishing East European and American Interests

The political leaders of Eastern Europe play to American fears about generalized instability as a way of increasing support for their early admission to the alliance. They also stress, sometimes to an implausible extent, the importance of their region to the United States. Czech president Vaclav Havel, for example, contends that "it is in the vital interest of the democracies to support freedom and democracy elsewhere in Europe and even to defend it. If they rejected such a role, they would clearly put themselves at risk."[6] In a statement clearly directed at U.S. officials and the American public, he says:

> I am convinced that the American presence in Europe is still necessary. In the 20th century, it was not just Europe that paid the price for American isolationism; America itself paid a price. The less it committed itself at the beginning of European conflagrations, the greater the sacrifices it had to make at the end of such conflicts.[7]

The West would "find once more that when Prague, Warsaw or Budapest is at stake, the fate of freedom on this planet is at stake with them." Conversely, "if a situation is achieved in which Europe will never again drag the rest of the world into war and instead will radiate peace and tolerance," the United States and the world will clearly benefit.[8]

Most political leaders have an unfortunate tendency to regard their own countries as the center of the geostrategic universe, and as Havel's soaring, universalist rhetoric indicates, he and his East European colleagues are no exception. What is less understandable is the willingness of some American leaders to attach similar importance to the policy preferences of the East Europeans. When asked at a press conference in June 1993 if the United States should remain in NATO, President Clinton responded:

> The clearest example that I know to give you that NATO is not dead was provided by the leaders of the Eastern European countries. . . . Every one of those presidents said that their number one priority was to get into NATO. They know it will provide a security umbrella for the people who are members. And I think we should continue to be involved in it.[9]

Clinton's comment was a worrisome example of letting other nations define America's security interests. The issue should not be whether current or aspiring protectorates of the United States desire U.S. security guarantees. It is hardly a startling revelation that nations living in a dangerous international neighborhood want a friendly superpower to provide some insurance against aggression—especially when they would have to pay little or nothing for that insurance. The history of international relations is replete with examples of small, vulnerable countries seeking out distant protectors to neutralize an actual or potential threat posed by a stronger nearby state.

The situation in Eastern Europe fits that classic pattern. There, a band of relatively weak states is sandwiched between two larger and more powerful states—Russia and Germany. Moreover, both of those great powers have previously engaged in expansionist actions at the expense of their neighbors. The apprehension of the nations in that region is entirely understandable, even if there is no looming threat to their security at the moment. Since they cannot be certain that the present benign situation will persist, trying to gain shelter behind the shield of a U.S.-led NATO makes perfect sense from the standpoint of their self-interest.

The pertinent question for U.S. policymakers, however, should be whether incurring the costs and risks of providing such protection really benefits the United States. When President Clinton blithely assumes that NATO is worthwhile because the East Europeans want to join, and that the United States should stay in the alliance for that reason, he subordinates America's interests to the interests of other nations. The wishes of European governments should not be a relevant factor, much less the determining factor, in making decisions about U.S. policy toward either the current version or an expanded version of NATO.

Georgetown University professor Earl C. Ravenal has warned trenchantly that U.S. policymakers have a habit of "adopting every country's threats" as America's own.[10] Ravenal's observation is especially applicable to the proposals to enlarge NATO. The more members the alliance takes in, or even the more territory that comes under its security jurisdiction, the greater the number of quarrels, conflicts, and adversaries NATO's current members will encounter. With every additional member (or protectorate), the alliance will become responsible for resolving disputes involving that party or defending it from its enemies. Author William Pfaff inadvertently underscores the breathtaking extent of such responsibilities when he suggests that "acting through NATO, the Western powers could guarantee all Central and East European frontiers against non-negotiated change and against external threat."[11]

Such obligations would have direct and disturbing implications for the United States as NATO's acknowledged leader. Conflicts and rivalries that are currently of little or no concern to the United States would, by definition, become relevant with the expansion of the alliance. That change could easily bring America into confrontations with nations (or political factions within nations) locked in armed struggles, even though those parties presently have no reason for hostility toward the United States.

Making every nation's threats our threats is especially unwise in Eastern Europe, given the number of actual or potential conflicts. Americans ought to be more than a little nervous when Vice President Al Gore asserts, "It is time for NATO to address Europe's new security challenges, such as consolidating democracy's gains among NATO's eastern neighbors and warding off ethnic conflicts."[12] Columnist Patrick Buchanan cites the danger of such a mission concisely: "If the United States extends NATO security guarantees to Eastern Europe, the United States will be committed to go to war in perpetuity to freeze in place a balance of power everyone knows cannot long endure. That is a prescription for war."[13] Attempting to shield the East European states from Russian domination is by no means the only risk factor. Indeed, that mission may not even be the most probable source of danger. There are plenty of unresolved problems among the other Central and East European countries, and any one of those disputes could explode into armed conflict.

## Sources of Turmoil

Hungary, one of the nations most prominently mentioned for inclusion in NATO, has ethnic problems on three fronts. One of those problems involves another Visegrad country and prime candidate for NATO membership, Slovakia. While the NATO summit sessions were taking place in January 1994, leaders of the 600,000 Hungarians living in Slovakia held an assembly of 3,000 delegates to demand self-government. The gathering created deep fears among Slovaks that the Hungarian minority—some 11.5 percent of the country's population—had begun a movement to secede. Interestingly, the successful Slovak secessionist bid, which led to the dissolution of Czechoslovakia in 1993, began in much the same fashion. "Most Slovaks understand this as a first step to declaring an independent region that will later be joined to the Hungarian republic," Slovakia's president, Michal Kovac, stated in an interview.[14]

Secessionist sentiments could have serious consequences for relations between Slovakia and Hungary. It is unsettling that the Hungarian government has thus far refused its neighbor's request for a written assurance that it will not seek to alter their border. Budapest's official explanation is that such an agreement is unnecessary because Hungary is already a party to the Helsinki Final Act of 1975, which prohibits any border changes by force. (Bonn's similar attempt to evade Poland's request for a bilateral accord on the German-Polish border on the eve of Germany's reunification sparked fears throughout Europe of German revanchism.) Slovakia is noticeably uneasy about Hungary's intentions. Slovak leaders recall that present-day Slovakia was under Hungarian rule for several centuries before World War I. More recently, Slovakia's southern border region—the same area now largely inhabited by ethnic Hungarians—was given to Hungary by Nazi Germany during World War II. Slovak apprehension was exacerbated in 1990 when then Hungarian prime minister Jozef Antall declared that he represented the interests of all Hungarians living abroad as well as those living in Hungary itself.[15]

Some 2 million of that Hungarian diaspora live in the Transylvanian region of Romania. As did Hungarians in Slovakia, they came under foreign rule with the collapse of the Austro-Hungarian empire after World War I and the arbitrary borders established by the victors in the 1920 Treaty of Trianon. Budapest's concern about the treatment of Hungarians living in Transylvania, therefore, long predates the breakup of the Warsaw Pact. Relations between Budapest and Bucharest have been frosty at best since the end of the Cold War, and the treatment of Romania's Hungarian population has been the principal problem. Hungary is angry that Romania's long-standing practices of excluding ethnic Hungarians from government employment, discriminating against Hungarians who seek admission to universities, and restricting even the most mundane Magyar cultural practices have not abated.[16]

An even more serious dispute is percolating with Serbia because of the increasingly precarious status of the 400,000 Hungarians in the Serbian province of Vojvodina. Belgrade has tightened its control of Vojvodina in recent years—and of the predominantly Albanian province of Kosovo—and the Hungarians are understandably nervous. The campaigns of ethnic cleansing conducted by Serb forces in portions of Croatia and Bosnia and the crescendo of chauvinistic rhetoric in Serbia have done little to ease that worry.

Hungary may have increased the danger to its ethnic brethren by helping to enforce the UN economic embargo against Serbia and by allowing NATO planes to use Hungarian airspace to monitor the no-fly zone over Bosnia. Both actions were apparently taken to demonstrate Hungary's value and loyalty to the Western alliance, thereby enhancing the country's prospects for eventual membership, but they have seriously damaged relations with Belgrade.[17]

Despite Budapest's desire to ingratiate itself with NATO, the limit was reached in February 1994 when alliance leaders sought permission for AWACS aircraft to use Hungarian airspace to coordinate air strikes on Serb positions in Bosnia. The Hungarian government rejected that request, specifically citing the potential danger of retaliation against Hungarians living in Serbia. "Hungary will not take part," confirmed Prime Minister Peter Boross. "We have to live with the Serbian people for hundreds more years."[18] It was troubling, though, that Hungarian officials hinted that the answer might have been different if NATO had offered their country membership instead of the vague Partnership for Peace—or at least had given an explicit pledge to protect Hungary from Serb retaliation. One bitter Hungarian diplomat complained that NATO officials "gave the impression they expected us to allow them to fly AWACS over Hungary for the airstrikes. . . . Yet when we ask for a decent Partnership for Peace, the door is only half open."[19] The incident suggests both the extent of the possible ethnic entanglements if NATO offers membership to Central and East European states and how membership might embolden those states in their confrontations with neighboring countries—with NATO expected to pick up the pieces.

The problems involving Hungary, Slovakia, and the other Visegrad countries, however, pale in comparison with the disputes brewing elsewhere in the region. Secessionist pressures continue to simmer in the predominantly Russian transdniester region of Moldova. At the same time, some Romanian expansionists hope to ultimately absorb Moldova into a greater Romania. That is not an unrealistic goal, since most of Moldova was the Romanian province of Bessarabia during the 1920s and 1930s; the Soviet Union annexed the region during the early stages of World War II—one of the territorial dividends of Stalin's brief partnership with Hitler.

The newly independent republic of Macedonia occupies an even more precarious position. Macedonia is the potential arena for conflict born of the competing territorial claims of several neighbors.[20] Bulgaria qualified its reluctant recognition of Macedonia's independence by reiterating its long-standing claim that Macedonians are merely "western Bulgarians." Greece refuses to extend even conditional recognition. Athens vehemently objects to the new republic's use of the name "Macedonia," contending that it implies territorial claims on the Greek province of Macedonia. Greece was so agitated by Skopje's appropriation of "Hellenic symbols" (including the 16-point "star burst" pattern in Macedonia's flag) and provisions of the republic's constitution that seemed to contemplate a "greater Macedonia" that Athens imposed a trade embargo on its northern neighbor.[21] Since most of Macedonia's commerce with the outside world, including vital supplies of energy, flows through the Greek port of Thessalonika, the virtual closure of the Greek-Macedonian border has had serious economic consequences for the new republic.

Macedonia's troubles hardly end with the reluctant tolerance shown by Bulgaria and the exaggerated accusations of Greece. Serbian nationalists note that Macedonia was the southernmost portion of the medieval Serbian kingdom before it was conquered by the Ottoman Turks in the 14th century. Such historical claims might be enough to interest Belgrade at some point. It is also an ominous development that Serbian leaders now claim that there are 400,000 Serbs living in Macedonia—some 20 percent of the population.[22] That number is far larger than the 44,000 recorded in the last census in 1991. Even if the census count understated the size of the Serb population—and there is some evidence that it did—most outside observers believe that the more recent claim is wildly inflated. One must wonder whether the motive of Serb leaders in both Serbia and Macedonia is to create the foundation for a "minorities problem" that could then be used to justify military action against the Macedonian government.

Some Macedonian officials worry that their country could become the victim of joint machinations by Greece and Serbia. The more extreme versions of such conspiracy theories, which assume that the two countries would invade and partition Macedonia, are probably unwarranted. But the more general fear is not irrational, given the close ties between Athens and Belgrade. Greece has clearly taken steps that have the effect of destabilizing Macedonia, and Serbia shows no signs of remorse at that development.

A more immediate worry for Skopje, though, would seem to be Albania's maneuvers. Albania has demanded that Macedonia's Albanian population (21 to 30 percent of the total, depending on the source of the estimates) be granted political and cultural autonomy. Tirana apparently has territorial designs of its own on the predominantly Albanian portions of Macedonia, and relations between the two countries have grown progressively more frayed. In November 1993 Macedonian officials claimed to have foiled a secessionist conspiracy directed by the Albanian military command.[23] Tensions rose further in February 1994 when the political party representing the interests of Macedonia's Albanian community split, with radicals repudiating the theretofore moderate leadership and moving to establish closer links to Tirana. Macedonia's leaders now state candidly that the status of the country's Albanian minority and relations with Albania constitute the biggest problem for their country.[24]

None of those developments is taking place in a historical vacuum. Macedonia was the principal arena of and the principal prize for two brief but bloody conflicts earlier in this century: the Balkan wars of 1912 and 1913. Those wars involved many of the same countries that are now adopting such hostile policies toward the new republic.

The United States and its NATO allies are wandering into a potential geopolitical maelstrom without much apparent thought of the dangers. There are credible reports that both Albania and Macedonia have become bases for U.S. intelligence operations in Serb-held territory in Bosnia and in Serbia itself. (Those include spy plane flights from Albania's Gjader Air Base.) High-level Western military officials, including Gen. George Joulwan, NATO's supreme commander; Gen. John Shalikashvili, chairman of the Joint Chiefs of Staff; and Gen. Robert Oaks, commander of the U.S. Air Force in Europe, conferred with Albanian leaders in early 1994.[25] Albania's president Sali Berisha is clearly enthusiastic about the rapidly expanding relationship with NATO, vowing, "If NATO needs our facilities, NATO will have them."[26] His government was also among the first to join the Partnership for Peace.

The NATO, especially U.S., connection with Macedonia is growing as well. Macedonia's interior minister, Ljubomir Frchkovski, boasts of a "significant" American presence, beyond the 500 "peacekeeping" troops that Washington acknowledges. Hinting at U.S. intelligence activities, he told reporters, "I'm sure you'll understand if I don't go into details."[27]

NATO's fixation on stopping further Serbian expansion by establishing a military presence in Albania and Macedonia could cause the alliance to be blind-sided by conflicts that erupt from other sources. The UN peacekeeping force in Macedonia, largely composed of troops from the United States and the Scandinavian countries, could easily become entangled in a dispute between Tirana and Skopje, given the conflicting agendas and deteriorating relations between those two governments. Washington's increasingly cozy relationship with Albania might also encourage Berisha and his followers to foment trouble among Albanians in Kosovo as well as in northwestern Macedonia. The complexity of politics in the Balkans underscores the point that military commitments that have one purpose and are focused on one objective can nevertheless easily produce disastrous unintended consequences.

Although NATO leaders speak casually of an alliance mission to head off conflicts in Eastern Europe, the Macedonian situation offers a glimpse of the many sources of trouble and the extreme volatility of the regional political environment. If NATO partisans believe that the stability mission is a relatively low-risk venture, they should disabuse themselves of that notion.

An enlarged NATO would face three types of dangers. One would be the obligation to defend a signatory from aggression by an outside power. The more expansive the alliance's security jurisdiction, the greater the potential for having to face down or wage war against some member's regional nemesis. That problem is all the more troublesome because aggression can be an ambiguous concept. In some cases aggression might be obvious, but in others it might be difficult to determine who was victim and who was perpetrator. Indeed, some states might even see NATO membership as a shield behind which they could settle old scores or pursue expansionist agendas. Although in theory the treaty signatories would have no obligation to rescue a member that committed aggression against another country, it would not necessarily be easy to sort out the truth in the midst of a crisis.

Moreover, any waffling on article 5 obligations would threaten to undermine the credibility of the alliance's entire security structure. Recognition of that danger would create additional pressure to rescue the beleaguered party even if there were doubts about its status as an innocent victim. Such matters may have little relevance while NATO's membership is confined to the stable, democratic, status quo nations of Western Europe and North America, but they would become the source of far greater concern for an alliance that included the less predictable and possibly revisionist nations of Eastern Europe.

Although some proponents of enlargement contend that comprehensive expansion of NATO would help alleviate the potential conflict between "ins" and "outs," that belief is largely an illusion. An alliance that included Russia and the other states of the former Soviet Union as well as the nations of Central Europe might overcome the danger of again dividing Europe into rival security blocs, but it would not solve the other problems associated with NATO's expansion. Unless NATO became congruent with the United Nations, there would always be some nations included in the fold while others were excluded. It is equally certain that some alliance members would have enemies among the nonmembers. The underlying treaty obligations and the risks they entail would remain qualitatively the same.

The second source of danger is the possibility of a war between NATO members. A favorite argument of proponents of NATO's expansion is that a larger alliance would internalize potential conflicts and thereby increase the probability of a peaceful solution. They contend, for example, that Greece and Turkey might well have gone to war on several occasions—especially in 1974 when Turkish forces invaded Cyprus—if it had not been for their mutual NATO membership.

Political pressure from treaty partners might induce feuding members to settle a dispute before resorting to force, and the existence of institutional mechanisms of conciliation could serve to dampen volatile quarrels in some cases. There is no guarantee, however, that such constraints would always prove sufficient. The result of such a "deterrence failure" would be a war between NATO members, and Atlanticists ought to realize that even one such episode might be enough to shatter the alliance.

Indeed, the Greco-Turkish case may not offer as much comfort as some NATO enthusiasts believe. The fact that each country has frequently contemplated war against its nominal ally ought to be reason for concern, not complacency. (Monteagle Stearns, a former U.S. ambassador to Greece, contends that the United States and other NATO members had to step in on three occasions—in 1964 and 1967 as well as in 1974—to prevent the two countries from coming to blows.)[28] It is especially worrisome that Athens and Ankara could repeatedly approach the brink of war even though such action would have disrupted NATO unity and increased the security menace posed to all members by their common adversary, the Soviet Union. Those who believe that the mere fact of membership in NATO dissuaded the Greeks and Turks from engaging in intra-alliance bloodletting ought to ask themselves how confident they are that either party would have shown the requisite restraint without the looming threat posed by Moscow.

Even such a strong proponent of U.S. activism and a continuing role for NATO as Stearns concedes that there are numerous factors that could trigger a war between Greece and Turkey in the post–Cold War era. The festering dispute over Cyprus is the most likely source of trouble, but he identifies others. They include quarrels over the exploration and exploitation of mineral resources of the Aegean shelf, territorial disputes involving a number of Aegean islands, and provocative deployments and maneuvers by Greek and Turkish naval forces.[29]

It is also revealing that proponents of NATO intervention in Bosnia routinely cite as a major justification the danger that otherwise the conflict will spread southward. That image of toppling geopolitical dominoes usually culminates in an armed struggle between Greece and Turkey, as both countries are drawn into a wider Balkan war.[30] Those who raise that specter implicitly have little confidence that NATO membership would prevent Greco-Turkish bloodletting.

In the absence of a powerful mutual adversary, it is merely an article of faith that NATO members would never come to blows. That is especially true if the alliance moves east into areas that harbor numerous unresolved conflicts.

In addition to the dangers entailed in striving to keep new members from fighting each other or in protecting a new member from a regional adversary, NATO could find itself drawn into internecine conflicts throughout the region. Once the alliance views stability in Eastern Europe as a pertinent, indeed vital, security issue, it cannot easily regard disruptive developments from any source with detachment. There would also be enormous pressure to support the incumbent government of a NATO member in putting down an insurgency that threatened the nation's territorial integrity.

Indeed, that already appears to be happening in Turkey. Ankara has sought to suppress a persistent guerrilla war being waged by the Kurdish minority in the mountainous southeastern region of the country. The scale of the conflict can be gauged by the offensive that the Turkish army waged in the spring of 1994, which involved some 130,000 troops. More than 11,000 Kurds have perished in the fighting that has occurred intermittently for more than a decade, and Amnesty International and other human rights organizations have condemned the Turkish government for numerous atrocities. Among other crimes, Ankara's forces have reportedly evacuated and destroyed some 900 Kurdish villages to prevent them from aiding the rebels—an act that would seem to constitute "ethnic cleansing."[31]

Nevertheless, the United States and other NATO members continue to supply Turkey's military with sophisticated weapons, which the *New York Times* observes caustically "were certainly not meant for use against Kurdish villages."[32] Such support has generated some controversy in the United States and even more in Germany, which is another important arms supplier. Domestic criticism even impelled Bonn to suspend weapons shipments in 1993, when Turkish atrocities reached egregious levels. Germany resumed arms shipments in May 1994, however, despite no noticeable improvement in the behavior of the Turkish military. Interestingly, Bonn's actions came after a steady stream of complaints by Ankara about the lack of support from its NATO allies.

## The Elusive Concept of Aggression

The stability issue could lead to NATO's entanglement in a variety of other internecine conflicts. For example, a renewal of fighting in Moldova as a result of the transdniester region's efforts to secede could prove troubling even if Russia and Romania attempted to stay out of the fray. Similarly, a rebellion by Macedonia's Albanian minority would arguably pose a security threat to the alliance, given the potential for destabilizing the country and drawing in Albania and other outside powers. The United States has already warned that any move by Belgrade to escalate its coercive practices against the Albanian majority in Kosovo would be a matter of grave concern and might well trigger a Western military response.[33] One motive for President Clinton's decision in the summer of 1993 to station U.S. troops in Macedonia as part of an international peacekeeping force was to discourage the Serbs from pursuing an ethnic-cleansing policy in Kosovo. One "high-level" administration official admitted to reporters that the administration had even considered sending U.S. forces into Kosovo as well as Macedonia to thwart further Serb coercion.[34]

Such incidents suggest that it is not always easy to make a clear distinction between cross-border and internecine conflicts. The fighting in Croatia and Bosnia, for example, has exhibited characteristics of both. Advocates of intervention have, of course, routinely accused Serbia of "aggression" against its neighbors. Former British prime minister Margaret Thatcher's reasoning is typical: "A sovereign state, recognized by the international community, is under attack from forces encouraged and supplied by another power. This is not a civil war but a war of aggression, planned and launched from outside Bosnia though using the Serbian minority."[35]

But allegations of aggression are entirely dependent on the decision of the United States, the members of the European Union, and the

United Nations to recognize the independence of Croatia and Bosnia over Belgrade's vehement objections. By declaring that the Yugoslav state was defunct, and disregarding Serbia's desires to keep the federation intact, the United States and its NATO allies arbitrarily redefined a civil war as one of external aggression. International relations scholars Robert W. Tucker and David C. Hendrickson note that under international law a better case could be made that the recognition of Bosnia's independence "constituted an illegal intervention in Yugoslavia's internal affairs, to which Belgrade had every right to object."[36]

Washington and the EU states insisted on regarding the boundaries of Croatia and Bosnia as thenceforth sacrosanct international boundaries. But those boundaries were purely artificial creations, imposed by communist dictator Josef Broz Tito shortly after he consolidated his power at the end of World War II. They were meant to be *internal* (virtually the equivalent of "provincial") lines of political and administrative demarcation within Yugoslavia, not the boundaries of independent, sovereign states. Those jurisdictions also had extremely weak historical roots and made little sense from the standpoint of ethnic distribution or economic relations. In particular, Tito's creation of such "republics" left large Serb minorities in both Croatia and Bosnia—a deliberate ploy on his part to dilute Serb political influence.

When Croatia and Bosnia declared independence, those Serb communities, fearing that they would be the targets of discrimination or even outright persecution, launched secessionist bids of their own. Howard University professor Nikolaos Stavrou contends that ill-considered action on the part of the United States and its European allies exacerbated an already dangerous situation. "With amazing haste, administrative and geographic borders had been converted to international ones without much concern for the ethnic makeup of these new entities. No serious consideration was given to the implications of recognizing new states prior to legally securing autonomy for ethnic groups within these states."[37]

For the NATO powers to insist that the Bosnian conflict is not a civil war but a case of international aggression—therefore warranting a NATO military response—because Bosnia became a recognized member of the international community in 1992 is reminiscent of a logical flaw that Abraham Lincoln exposed with his renowned caustic wit. "If you call a dog's tail a leg," Lincoln reportedly asked a political colleague, "how many legs would a dog have?" His colleague replied that it would then have five. "No," Lincoln responded, "calling a dog's tail a leg doesn't make it a leg." Calling the Bosnian struggle a case of cross-border aggression rather than a civil war does not make it one.

Moreover, even if one accepted the dubious proposition that the West's action in recognizing the independence of Croatia and Bosnia without changes in their presecession boundaries was legitimate, the conflicts in those two countries would still be hybrid internecine and cross-border wars. Although there is certainly extensive "interference" by an "outside power" (Serbia), most of the Serbs who are engaged in the fighting do not come from Serbia. The vast majority of the insurgents in Croatia are inhabitants of the predominantly Serb region of Krajina, and there are significant differences between their political agenda and that of the Milosevic regime in Belgrade. (Serbia is more amenable, for example, to a peace agreement that would accord the Krajina Serbs autonomy rather than full independence.)

Similarly, most of the "Serb forces" that have challenged the authority of the Muslim-dominated government in Sarajevo come from Bosnia, not Serbia. Belgrade assuredly influences their actions and provides its allies with significant logistical support, but it is an exaggeration to suggest that the Milosevic government tightly controls their actions. As in the case of the relationship between Serbia and the Krajina Serbs, there are noticeable political differences. Belgrade has repeatedly stated that it would be willing to endorse a peace settlement that would require Bosnian Serb forces to give up some of the territory they have conquered.[38] Needless to say, Radovan Karadzic and other Bosnian Serb leaders are less than enthusiastic about that proposal. Contrary to the simplistic conclusion of interventionists such as columnist William Safire, the Bosnian Serbs are hardly Belgrade's "puppet forces."[39]

Given the murky nature of the struggle in Bosnia, NATO's deepening involvement should not only be cause for concern, it should be an alarm bell warning of the dangers entailed in expanding the alliance. That war is precisely the type of difficult, convoluted entanglement NATO can expect if it embraces the mission of attempting to pacify Eastern Europe.

# 5. The Ominous Bosnian Model

The prospect of NATO's finding itself mired in messy cross-border or internecine struggles in Eastern Europe is not merely theoretical. NATO is already involved militarily in the Bosnian civil war, and there are indications that the scope of intervention will increase. Secretary of Defense William Perry confirmed that point in unusually candid terms in comments to reporters during a July 1994 trip to the Balkans.[1]

Bosnia may well be a prototype of what the United States and other members can expect if NATO expands eastward. Indeed, National Security Adviser Anthony Lake asserts that by attempting to resolve the war in Bosnia, "we help buttress the historic reforms underway in neighboring nations such as Bulgaria, Hungary and Romania—nations whose security is important to our own."[2] Moreover, "We have an interest in showing that NATO—history's greatest military alliance—remains a credible force for peace in this era." U.S. officials clearly do not regard NATO's Bosnian mission as an aberration.

On February 28, 1994, NATO forces engaged in the first combat operations in the history of the alliance as two U.S. Air Force F-16s shot down four Bosnian Serb planes for violating a no-fly zone that the United Nations had proclaimed. An even more serious incident took place on April 10 when NATO aircraft bombed Bosnian Serb artillery positions ringing the besieged Muslim city of Gorazde. The following day NATO planes launched a second strike, and President Clinton and other alliance leaders warned the Serbs that additional sorties might be ordered unless attacks on UN peacekeeping forces stationed in Gorazde ceased.

Those episodes were more than slightly ironic. NATO, an alliance created to prevent a Soviet invasion of Western Europe, had never fired a shot in anger during the long years of the Cold War, despite numerous tensions and several full-blown crises. With the Cold War over, alliance forces now found themselves engaged in combat in a murky civil war in the Balkans. *Washington Post* defense correspondent Barton Gellman captured the irony of the initial air engagement. "After preparing for decades to fight for survival against a powerful foe, NATO swooped down on a handful of inferior planes—home-grown Yugoslav technology from the 1960s—and dispatched them with what [an Air Force officer] called 'almost unsportsmanlike' efficiency."[3] It is unlikely that George Marshall, Dean Acheson, Ernest Bevin, and the other architects of NATO had such a mission in mind when they negotiated the North Atlantic Treaty.

NATO's military involvement in the Bosnian struggle has evolved incrementally. The first step took place in July 1992 when the naval vessels of several alliance members were sent to the Adriatic to help enforce the UN economic embargo against the rump Yugoslav state of Serbia and Montenegro. Five months later a more significant step occurred when the United Nations, at the urging of the United States, Britain, and France, proclaimed a no-fly zone over Bosnia, patterned after a measure that the Security Council had adopted to protect Kurdish inhabitants in northern Iraq from Saddam Hussein's wrath after the Persian Gulf War.

Enforcement of the Bosnian no-fly zone was virtually nonexistent during the succeeding months, as NATO failed to respond to hundreds of violations (primarily helicopter reconnaissance and resupply missions by the Serbs). The lack of enforcement, despite periodic verbal warnings from various NATO capitals, gradually made the no-fly zone the subject of mounting ridicule throughout the international community. But the no-fly policy was a sword of Damocles hanging over the necks of the Serbs. There was always the possibility that NATO would take military action, as it ultimately did.

Shortly before the downing of the Bosnian Serb planes for violating the no-fly zone, the NATO powers took another step that increased the probability of deeper intervention. On February 4 a mortar round exploded in a crowded marketplace in Sarajevo, killing 68 people. Although the UN command stated repeatedly that it could not determine which faction had fired the fatal shell, the United States and its allies operated on the assumption that the Serbs were responsible. Consequently, NATO leaders gave the Serb forces an ultimatum to withdraw their tanks and artillery pieces at least 12 miles from Sarajevo and cease shelling the city within 10 days or be subjected to air strikes.

The Serbs complied with that ultimatum (after the deployment of Russian peacekeeping forces in and around Sarajevo), and an immediate crisis was averted. But the episode had far-reaching implications. The level of NATO involvement in the Bosnian conflict escalated significantly, for the alliance had now become the military protector of the Bosnian Muslim government's capital city. Predictably, advocates of full-scale Western intervention demanded that NATO extend the Sarajevo ultimatum to protect Gorazde, Tuzla, Srebrenica, and other Muslim enclaves in eastern Bosnia that were under siege by Serb forces. The United States initially resisted such demands; as late as April 3, Perry stated that there was no intention of using NATO air power against the Serb guns around Gorazde.[4] Ultimately, however, Washington acquiesced to the wishes of France and other alliance members who favored a more assertive policy and authorized air strikes.

It should not come as a great surprise that NATO shows signs of drifting into the Bosnian war under the guise of an international peacekeeping operation. As early as June 1992 there were indications that NATO leaders were willing to involve the alliance in peacekeeping (or more accurately, peacemaking) missions in Bosnia-style conflicts. Officials at the foreign ministers' conference in Oslo adopted a resolution explicitly making troops available for future peacekeeping or peace-enforcement operations approved by the Conference on Security and Cooperation in Europe (CSCE).[5] If the alliance was willing to place its forces at the service of the untried CSCE, there was little doubt that it was prepared to place them at the service of the United Nations.[6] There may be some question whether NATO is acting as the agent of the United Nations or vice versa in Bosnia, but the relationship has been a close, if sometimes turbulent, one throughout.

A cynic might conclude that the NATO powers are merely using the United Nations as a convenient institutional cover to legitimize political and military intervention. But the relationship is more complex. UN officials have asserted noticeable decisionmaking authority, sometimes to the irritation of the United States and its allies. That has been especially true in connection with various operational matters. Secretary General Boutros Boutros-Ghali and his colleagues, for example, insisted that the United Nations, not NATO, had the authority to order air strikes. There were several occasions, most notably in implementing an ultimatum to Serb forces besieging Gorazde in April 1994, when NATO military leaders wanted to order additional attacks but were overruled by UN officials in Bosnia or New York.[7]

NATO's serving (at least theoretically) as an agent of the United Nations in armed conflicts creates other problems and dangers in addition to those that would exist if the alliance were acting purely on its own authority. In the latter case, NATO would at least have control over military strategy and it would have the right to withdraw from the enterprise if it was not going well. With the involvement of the United Nations, that control is diluted. There is no guarantee that the alliance would be able to use the requisite force that its own military and political leaders deemed essential to carry out the mission. In fact, there is no certainty that NATO would be able to take the steps necessary to protect its troops in combat environments without getting the blessing of UN bureaucrats—a prospect that understandably worries top military commanders in the United States and other alliance countries. Important political decisions, such as those to opt for either escalation or withdrawal, might also be contingent on UN approval.

NATO's assumption of pacification missions in Eastern Europe on its own authority is a disturbing enough development. Doing so at the behest of the United Nations (or perhaps down the road at the behest of the CSCE) is more alarming. As on so many other points, the Bosnia operation provides a glimpse of what can be anticipated from such a relationship.

## Washington's Deepening Involvement

At the same time that the scope of NATO's military engagement in Bosnia deepened in 1993 and 1994, key alliance members, especially the United States, became increasingly entangled in that country's underlying political quarrels. By February 1994 Washington had adopted a high-profile diplomatic role in an effort to resolve the conflict. The centerpiece of that effort was a move by Clinton administration officials to orchestrate a political settlement between Bosnia's warring Muslim and Croat factions. To the surprise of many both inside and outside Bosnia, the administration appeared to achieve its objective. It was less certain, however, whether the apparent diplomatic coup accomplished anything worthwhile. Indeed, the bulk of the evidence suggested that the accord actually exacerbated the situation by creating new conditions for instability and conflict.

U.S. officials stated that the medium-term objective was to establish a Bosnian Muslim-Croat state in a loose confederation with Croatia.

Some officials even hoped that the Bosnian Serbs could ultimately be enticed or coerced into joining that new political entity as part of a comprehensive settlement.

The U.S. strategy was dubious for several reasons. Resuscitating the Muslim-Croat alliance against the Serbs merely recreated the conditions that caused the Bosnian civil war. Indeed, administration policymakers conceded that the purpose of the confederation plan was to isolate the Serbs and force them to make concessions. Anthony Lake stated bluntly: "The agreement between the Muslims and Croats is of strategic consequence. It changes the power equation in the area and places greater pressure on the Serbs to join negotiations on their future status in Bosnia and on territorial issues."[8]

The principal Serb concession that Washington demanded was to relinquish enough land to give the new Muslim-Croat state control of at least 51 percent of Bosnia's current territory. At that point in the war, Serb forces controlled approximately 70 percent. U.S. officials—echoed by most of the American press—incessantly cited the 70 percent figure as evidence of the extent of Serb "aggression." Those who made that assertion acted as though no Serbs had lived in Bosnia before the war—as though they dropped in from some other planet and seized 70 percent of the land. The reality was quite different. Although Serbs constituted only 33 percent of Bosnia's prewar population, they were a largely rural people. Consequently, some 60 percent of Bosnian territory was inhabited by Serbs before the first shot was fired in the civil war.

Washington's policy in essence demanded that the Serbs accept a settlement that would give them less land than they had possessed at the start of the conflict, despite their victories on the battlefield. Even if the threat of NATO military intervention intimidated the Serbs into making such a humiliating capitulation, it would be an empty accomplishment for American or NATO diplomacy. The boundary between the new Muslim-Croat federation and a Bosnian Serb state (whether or not officially recognized by the international community) would become one of the world's perennial flash points. Serbs would nurse grievances about the territories "stolen" by the Croats and Muslims— with the aid of the United States—and exploit every opportunity to undermine the settlement. Such conditions would hardly create the foundation for an enduring peace.

In addition, the proposed Bosnian Muslim-Croat state—agreed to by representatives of the two factions at a signing ceremony at the White House on March 18—was a congenitally unstable political monstrosity. It was replete with arrangements (such as a presidency that will rotate annually between Muslim and Croat politicians) reminiscent of the awkward balancing measures that Lebanon used in a vain effort to preserve peace between its Christian and Muslim factions. The collapse of the Lebanese state into a bloody civil war in the mid-1970s offers little reason for optimism that such a strategy will work any better in Bosnia.

Washington appeared to harbor the illusion that the Muslims and Croats, because they initially cooperated against the Serbs, had basically compatible objectives. But they did not. Although both groups supported Bosnia's secession from an increasingly Serb-dominated Yugoslavia, they did so for very different reasons. The Muslims, as the most numerous faction in Bosnia, wanted an independent state in which they would be dominant. The Croats, as the least numerous group, had little enthusiasm for that outcome. They regarded Bosnia's independence as merely an interim step toward a merger of the predominantly Croat portion of the country with Croatia.

Two factions with such incompatible goals are unlikely to coexist for long in the same country. If, under the federation proposed by United States, the Muslims eventually sought to exercise their power as the majority, the Croats would want out of the arrangement. Moreover, they would probably ask Croatia for protection, a move made easier by the loose, ill-defined "confederation" with that country that Washington designed for the new Bosnian state. One could scarcely imagine a more effective blueprint for mischief, instability, and conflict.

The fuzzy-minded nature of U.S. diplomacy did not end with the creation of the new Croat-Muslim federation. A senior official indicated in mid-March that the administration still strongly favored keeping Bosnia intact by persuading the Serbs to join the federation. Although he did not rule out the possibility of recognizing a secessionist Serb state, the conditions under which such a state would be recognized were daunting. Indeed, U.S. policy apparently gave the Sarajevo regime a virtual veto over that option. The senior official stated, "If, at some point, the Bosnian Government tells us, 'Look, this is the best chance for our entity,' we'd have to take that into consideration." A second anonymous high-ranking official expressed a similar perspective. "It will be very difficult if the Serbs want to secede. They would have to give much more territory than if they remain within a loose union in Bosnia. But our bottom line is that *if eventual secession is acceptable to the Bosnian Muslims,* then we will support it."[9]

The U.S. decision to pressure three mutually antagonistic ethnic factions to live together in one country is ill advised. Bosnia's prospects for political viability were not promising even before the civil war erupted. After more than two years of bloodletting, the notion of a stable, united Bosnian state was a utopian fantasy. Just how fanciful were U.S. schemes for creating that entity became apparent when reporters began to press Clinton administration leaders about the specifics. The ever-helpful "senior official" outlined U.S. views.

> What we have in mind is that the central government of the whole would be weak, but the Muslim-Croat part would be stronger. The links to Croatia on the outside could be stronger than those to the Serbs within the country of Bosnia. You'd end up with an asymmetrical federation in Bosnia.[10]

It is more likely that one would end up with a colossal mess. The notion of a country in which two population groups have stronger political ties to an outside power than they do to each other is, to put it charitably, peculiar. It also raises the question of whether the Serb portion of the Bosnian triangle would have stronger ties to Serbia than it would to the Muslim and Croat communities in Bosnia. The implication of the official's thesis is that it would. Such a hopelessly convoluted political entity might be called many things, but "viable country" is not one of them.

Americans had reason to be concerned about such strange diplomatic machinations, since the administration had repeatedly indicated that the United States would provide troops as part of a UN-NATO peacekeeping force to implement a Bosnian peace agreement. Lake admitted that "we must be prepared for the possible deployment of NATO troops, including U.S. troops, to enforce the peace. Active American support would be essential."[11] Although Lake stated that a settlement would have to be "viable" before such peacekeeping forces were deployed, neither he nor any other administration official has described in detail what standards would be used to evaluate the viability of a peace accord. Given the mutual animosity exhibited by all three factions in the struggle, and the failures of numerous cease-fires, the duration of a peace accord might well be measured in months or even weeks. That is especially true of any settlement that does not reflect battlefield realities but is instead the product of intense pressure from the NATO countries or the United Nations.

The extent of the alliance's military involvement contemplated by the Clinton administration is clearly quite large. The U.S. contingent in a NATO peacekeeping force in Bosnia would go far beyond a token troop presence. Lake emphasizes that "we must bring our forces to bear in sufficient mass to get the job done. If our forces are deployed to Bosnia, they will go in strong. They will be part of a NATO force, not a UN force."[12]

The last comment reflects a supposed lesson from the debacle of the U.S.-UN intervention in Somalia. When 18 Army Rangers perished in a firefight with units of Somali clan leader Mohammed Farah Aideed in Mogadishu on October 3, 1993, critics in the United States were especially angry that the UN's cumbersome command structure had caused serious delays in sending reinforcements to rescue the beleaguered Americans. Congressional leaders and other influential players put intense pressure on the administration never to place U.S. troops under direct UN command again in combat situations. Clinton tacitly acquiesced to those demands; in conjunction with his decision to extend the U.S. mission in Somalia to March 31, the president emphasized that the remaining American troops there, as well as the additional units being sent to bolster their security, would operate under U.S., not UN, command.

Lake's comments confirmed that American policymakers had learned only the most limited and superficial lessons from the experiences in Somalia. Administration leaders apparently assume that all would have gone well with the U.S. intervention in that country if only Washington had insisted on a command structure not under UN control and if enough troops had been kept on the scene to intimidate the Somalis. Hence, Lake's insistence on U.S. participation in a NATO rather than a UN force in Bosnia and the assurance that the force would "go in strong."

But the failure of the U.S. mission in Somalia was attributable to more fundamental causes than a faulty command structure and a premature draw-down of forces in the spring of 1993. Although the UN (largely U.S.) intervention was initially portrayed as a humanitarian mission, there was always an underlying political agenda. UN officials stressed early on that the ultimate goal was the reconciliation of all Somali factions—basically, a nation-rebuilding project. The Bush administration in its waning weeks sought to keep the United States detached from that agenda, but the Clinton administration signed on to it with apparent enthusiasm.

The nation-rebuilding objective was unrealistic. Not only did it ignore the wishes of the people of northern Somalia, who had already proclaimed the secessionist Republic of Somaliland and wanted no part of a program of national reconciliation, but it soon ran afoul of domestic politics elsewhere in the country. UN officials (including U.S. special envoy Adm. Jonathan Howe) sought to marginalize the warlords and help create a new civilian leadership. It was that meddling in Somalia's internal affairs that impelled Aideed to regard the United Nations, and the U.S. forces that were helping to implement UN policy, as an enemy and to respond accordingly. Attacks by his militia forces drew the U.S. troops ever deeper into combat operations.

It is that parallel, not merely the supposed lessons about command structures and force levels, that should warn American leaders against encouraging a NATO peacekeeping mission in Bosnia. Despite President Clinton's assertions that NATO is not taking sides in the Bosnian conflict and seeks only to bring the fighting to an end, the Serbs have reason to view matters differently. Although the arms embargo imposed by the UN Security Council in September 1991 on the rapidly fragmenting Yugoslavia has worked to the disadvantage of the Muslims in Bosnia (the Serbs had most of the weapons from the old Yugoslav army), other actions have even more clearly damaged the Bosnian Serb cause. In particular, the United States and the other NATO powers have sided with the Muslim government throughout the conflict. They have not only oversimplified an exceedingly complex civil war by branding the Serbs as aggressors and portraying the Muslims as innocent victims, but they have taken numerous substantive steps against Serb interests.

NATO's enforcement of the no-fly zone effectively nullified Serb air superiority. Similarly, the creation of safe havens in eastern Bosnia thwarted Serb offensives that were poised to overwhelm several isolated Muslim enclaves. Washington's diplomatic initiative contemplates an accord that would not only erase Serb territorial gains achieved during the war but would support many of the political demands of the Muslim-dominated Bosnian government. Even before the Sarajevo ultimatum and the bombing of positions around Gorazde, the Serbs had every reason to view the notion of NATO's impartiality with a mixture of incredulity and cynicism.

Under such circumstances, it would hardly be surprising if Serb forces began treating UN peacekeeping troops from NATO countries—and the planes that give them air cover—as the adversaries that they are. A few days after NATO extended the Sarajevo "exclusion zone" concept to cover Gorazde and other Muslim enclaves, there was a nasty clash between peacekeeping troops and Bosnian Serb forces, with numerous fatalities. Even though the peacekeepers were technically UN troops, it is likely that the Serbs noticed that they came from Denmark—a NATO member.[13] NATO's protection of Gorazde and other safe havens has had another effect that is likely to anger the Serbs. As Sen. John McCain (R-Ariz.) concluded: "The Serbs interpreted our action as involvement on the side of the Muslims. So did the Muslims."[14] It did not take long for McCain's observation to be borne out. A correspondent for *U.S. News & World Report* soon reported, "That shield is now freeing Muslim forces to fight elsewhere."[15]

Even if the United States and its allies can somehow coerce the Bosnian Serbs into signing a peace accord against their interests and better judgment, there is little reason to believe that such an artificial, unstable agreement will end the fighting. A postsettlement NATO force in Bosnia is likely to be viewed as a symbol of occupation by the Serbs—and perhaps by disgruntled Bosnian Croats as well. NATO forces could easily become the target of retaliation just as U.S. troops did in Somalia—only on a larger scale.

It is questionable whether even the European members of NATO should incur the costs and risks of intervening in Bosnia, although they do have some interests at stake. The persistence of political turbulence on the borders of the European Union is undoubtedly unnerving. Nevertheless, a full-scale military intervention in Bosnia would be a high-risk undertaking with uncertain prospects for success. It would be reasonable for the West Europeans to conclude that the game was simply not worth the candle.

For the United States to contemplate intervention because of its leadership role in NATO is even less justified. Although the EU nations may have some legitimate security concerns in the Balkans, there is nothing at stake in the Bosnian conflict that remotely approaches being a vital American security interest. Only if one assumes that America's interests are congruent with those of the European members of NATO can a plausible case be made for U.S. military involvement.

### The Faulty 1930s Analogy

It has become a cliché among interventionists to compare Serbia to Nazi Germany, with Serbian president Slobodan Milosevic playing the role of the "new Hitler." According to proponents of U.S. military action in the Balkans, the United States risks a rerun of the tragic events of the late 1930s, culminating in a larger war, if it fails to stifle aggression in its early stages. At the very least, they warn, the Bosnian conflict will spread southward, drawing in Greece, Turkey, and other countries. Those who contend that today's situation in the former Yugoslavia is comparable to the wave of fascist aggression in the 1930s adopt a simplistic, rote interpretation of history.

The crisis in the 1930s involved one of the world's great powers—one with the second largest economy and a large, well-trained military force—embarking on a frightening and highly destabilizing expansionist binge. Serbia, on the other hand, has a population of 9.8 million (about the same as Belgium's) and a gross domestic product less than one-fifth of Denmark's. Indeed, even before the UN economic sanctions began to bite, Serbia's 1991 GDP of $18.75 billion was only modestly greater than Luxembourg's.[16] Belgrade's military forces, while not insignificant, consist largely of remnants of the old Yugoslavian federal army (augmented by the Serb militias in Bosnia and Croatia). The effects of the war and the UN arms embargo—despite some leakage—have combined to degrade the readiness of those forces. Although Serb military units might well be capable of mounting a ferocious resistance to an intervening army in Serbia itself, or in Serb-controlled portions of Bosnia and Croatia, they could not launch credible offensive operations against neighboring states, much less against the major industrial powers of Europe. We are not likely to see Serb armored divisions advancing on Paris or Serb invaders conquering Ukraine as part of a quest for *Lebensraum*.

In the late 1930s Germany was capable of creating a massive disruption of the international system; in the 1990s Serbia is capable only of modestly strengthening its position at the expense of its ethnic rivals within the boundaries of the former Yugoslavia. Not only does Belgrade not have territorial ambitions outside those borders, it lacks the economic and military power to pursue broader ambitions.

The other comparison to the 1930s made by proponents of U.S. and NATO intervention is equally flawed. The Serbs are like the Nazis in another way, interventionists contend: the rabid nationalism and ethnic hatred exhibited by the Milosevic regime and its followers. America and its allies cannot stand by, the argument goes, while "ethnic cleansing" takes place in Bosnia, or the West will again be passive accomplices to genocide.

Although that argument is more plausible than the notion of the Serbs as a serious strategic threat, it hardly justifies NATO involvement in the war. Serb forces in Croatia and Bosnia have undoubtedly committed serious human rights violations. Nevertheless, the so-called ethnic cleansing and other undesirable actions must be placed in perspective.

The ethnic-cleansing label itself is misleading as well as inflammatory. U.S. officials, aided by large portions of the Western news media, have sought to equate it with genocide. But as columnist Charles Krauthammer, MIT political scientist Daryl Press, and other skeptics point out, what is going on in Bosnia cannot accurately be termed genocide. Press notes that genocide has a very specific meaning: "The systematic annihilation of a racial, political, or cultural group."[17] Instead of exterminating members of other ethnic groups, however, the Serb objective has generally been to expel them from certain territories as part of an effort to create a "greater Serbia." Although that is certainly an odious practice—and is sometimes accompanied by acts of murder—it does not constitute genocide.[18]

Indeed, events in such places as Sudan and Rwanda more closely fit the definition. Nearly 1 million black Sudanese have perished at the hands of the Arab government in Khartoum during the past decade. Strife between Hutus and Tutsis in Rwanda in the spring of 1994 claimed the lives of at least 500,000 people—and perhaps as many as 800,000—almost all of them Tutsi civilians, over a period of three months. Estimates of fatalities—including those of military personnel—in Bosnia range from 115,000 to 200,000 in more than two and a half years of warfare. Yet most of the individuals who demand that the Western powers stop "genocide" in Bosnia are far less strident in their calls for military intervention in Rwanda, and virtually none of them contend that there is a moral obligation to intervene militarily in Sudan.

Contrary to the impression fostered by U.S. and West European officials, ethnic cleansing itself is hardly a practice unique to the Balkans.[19] American leaders should be especially circumspect about denouncing other countries for "cleansing" territories of indigenous inhabitants. It was not that many decades ago that the U.S. government expelled the various Native American tribes from lands that political leaders and their constituents coveted—often in violation of solemn treaties.

But one does not have to go back to the 18th and 19th centuries—a period that apparently constitutes "ancient history" in the minds of U.S. policymakers and, therefore, has no applicability to the Bosnian situation—to find pertinent examples of ethnic cleansing. One such incident occurred when the British colony of India was partitioned and granted independence in 1947. Millions of Hindus were expelled from the new Islamic nation of Pakistan while millions of Muslims had to flee predominantly Hindu India. Nearly 250,000 people perished in the bloodshed that accompanied the partition.[20] From the standpoint of U.S. culpability, an even worse episode took place at the end of World War II when nearly 16 million ethnic Germans were expelled from Poland, Czechoslovakia, and other East European countries solely because of their ethnicity. Not only did the United States fail to take any action to prevent that forced exodus, President Harry S Truman openly endorsed the step at the Potsdam Conference, with the sole proviso that the process be "humane."[21]

More recently, some 200,000 Greek Cypriots were driven from the northern portion of Cyprus when Turkish forces invaded and occupied nearly half of the island in 1974. Again, it is instructive to compare the passive NATO response to Ankara's actions with the alliance's periodic saber rattling about Serb policy in Croatia and Bosnia. There has never been the slightest consideration given to using military action against Turkey to restore the Greek Cypriots to their homes. The Serbs have a point when they contend that the United States and its allies employ a double standard.

Finally, advocates of a U.S.-led intervention in Bosnia also contend that even if Serbia did not pose a strategic threat and Serb practices were not morally odious, the conflict nonetheless would have great symbolic importance. Other aggressors around the world are watching how the West responds to Serb expansionism, interventionists contend. The editors of the *Wall Street Journal* expressed that argument succinctly.

> Bosnia is about more than Bosnia. Slobodan Milosevic is merely the irredentist of the moment. All over the world are pirates masquerading as national leaders, eager to invade and kill the people next to them under the guise of historic grievances. Saddam was the first post Cold War irredentist. China has been ceded Hong Kong: it wants the Spratleys. We know about Kim Il-Sung, Assad, Saddam (still), the bitter losers of the Russian empire, Aidid.[22]

Former deputy under secretary of defense Dov Zakheim offers a similar interpretation. "Only active intervention on the side of the Muslims can assure that Saddam, and other aggressors, will not conclude that his timing on Kuwait was simply off by a few years ... and that the rewards of aggression are there for the taking."[23] Former British prime minister Margaret Thatcher asserts bluntly, "Would-be aggressors are waiting to see how we deal with the Serbs."[24]

According to that thesis, a strong NATO response in Bosnia will deter other would-be aggressors. Bombing Serbia, the *Journal* insists, "makes sense, not because it will 'stop the fighting,' but as an act of deterrence." In the post–Cold War era, strategic thinkers must confront a dangerous world "in which the model of deterrence is no longer always a Clausewitzian war, a la Vietnam, but instead a tough but discriminate 'shot across the bow' of the sort used against the Barbary Pirates and the Bey of Algiers."[25]

Even if one accepts the simplistic notion that the fighting in Bosnia is a case of aggression by an irredentist "pirate," the policy prescription is fallacious. It rests on the dubious notion of deterrence by example. Such "indirect" deterrence is much more problematic than "direct" deterrence: confronting a specific expansionist power with a preponderance of force and a declared policy of using that force if certain acts are committed. Foreign policy analysts Christopher Layne and Benjamin Schwarz raise serious questions about the validity of deterrence by example. They note that arguments asserting the need for indirect deterrence contend that aggression even in intrinsically unimportant areas such as the Balkans must be resisted or disorder will spread to regions that are important to the United States and its allies. Such reasoning, they point out, bears a strong resemblance to the containment era's domino theory, and therein lies the problem.

> The domino theory ... has never reflected the real dynamics of international politics. Unlike the chain reactions posited by physics, in the world of statecraft crises are usually discrete

happenings—not tightly linked events. The outcome of events in potential hot spots like Nagorno-Karabakh, Moldova, the Baltics, Ukraine, Transylvania, and Slovakia will be decided by local conditions, not by what the United States does or does not do in the Balkans.[26]

Several recent developments support that thesis and create doubts about whether deterrence by example is a valid international relations concept. As Layne and Schwarz remind us, "Slobodan Milosevic was not deterred by U.S. action against Iraq; Saddam Hussein was not deterred by U.S. action in Panama; Manuel Antonio Noriega was not deterred by U.S. action in Grenada, Lebanon, or Vietnam; Ho Chi Minh was not deterred by U.S. action against North Korea; and Kim Il-Sung and Joseph Stalin were not deterred by U.S. action against Adolf Hitler."[27] Fareed Zakaria, managing editor of *Foreign Affairs*, makes a similar point. "The theory has to explain an embarrassingly large problem. If a demonstration of American force in one country chills the blood of would-be aggressors in another, why did the Persian Gulf War not deter the Serbians, Azeris, Sudanese, Georgians, and Somalis?"[28]

In light of the dismal record, those who contend that a strong NATO military response against Serb "aggression" will deter expansionist powers or factions elsewhere in the world have a difficult time making their case. Indeed, there is scant evidence that deterrence by example works even against recalcitrant regimes in the same region. If the theory were true, Noriega certainly should have gone to great lengths to avoid provoking the United States, given Washington's long-standing record of using force against small nations in Central America and the Caribbean. The invasion of the Dominican Republic in 1965 and the invasion of Grenada in 1983 served notice that it was dangerous to antagonize the United States. Nevertheless, for personal and domestic political reasons, the Panamanian dictator continued a confrontational course right up to the moment U.S. troops launched their assault in December 1989. Believers in indirect deterrence seem to be relying on faith rather than a rational theory supported by meaningful historical evidence.

The shopworn 1930s analogy to the contrary, the fighting in Bosnia is a parochial, albeit ugly, struggle with little importance outside the immediate region. It is a struggle over the territorial spoils resulting from the breakup of the Yugoslav federation. Although it may be an especially acrimonious political divorce, it need not and should not have wider strategic or moral significance. Even the worst-case scenario—the spread of the conflict to Kosovo and Macedonia, with subsequent intervention by such outside powers as Albania, Greece, Bulgaria, and Turkey—would not fundamentally alter that reality. Unless the United States recklessly puts its prestige on the line, and its military forces in harm's way, there is little intrinsic reason why a third Balkan war would threaten vital American interests any more than did the Balkan Wars of 1912 and 1913 (which involved many of the same parties).

### Making the 1914 Tragedy a Self-Fulfilling Prophecy

Today a Balkan war would pose less threat to U.S. interests than did the earlier struggles. In the years immediately preceding World War I, two increasingly antagonistic alliances confronted each other across the heart of Europe. Those rival alliances included all of Europe's great powers, and key members of both alliances were closely identified with Balkan clients. Thus, there was always the potential that a Balkan conflict would escalate to a continent-wide war (as ultimately happened in 1914) that could threaten important U.S. security interests.

Those who contend that even if the Bosnian crisis does not echo the late 1930s, it is a repetition of the events that led to World War I, misconstrue the reasons why a Balkan quarrel detonated that war. Columnist Paul Greenberg notes ominously, "If there is any doubt about the threat to world peace" posed by the fighting, "the dateline on many of these stories out of a dissolving Yugoslavia should be sufficient warning: SARAJEVO."[29] Similarly, *Wall Street Journal* foreign affairs correspondent George Melloan intones:

> June 28 is the 80th anniversary of the 1914 assassination of Archduke Francis Ferdinand, heir presumptive to the Austrian throne, in Sarajevo, by a Serbian nationalist. This touched off a fateful chain reaction that ultimately became World War I, arguably the most awful war in its utter mindlessness and slaughter that has ever been fought. By some lights, World War II was a sequel. All that from a very small beginning.[30]

Such simplistic analogies demonstrate that a little historical knowledge can be a dangerous thing. Sarajevo has no mystical significance, nor do the Balkans as a whole have inherent strategic or geopolitical importance. A crisis in the Balkans led to World War I not because of the region's intrinsic value (which was as minimal then as it is now) but because the major European powers foolishly identified their own vital interests with the outcome of its petty conflicts. As the RAND Corporation's Schwarz correctly observes, "The fuse for that war was lit in Sarajevo not because ethnic conflict existed in what is now Yugoslavia, but because great powers meddled in those conflicts."[31]

The situation today is considerably different. Europe is not cleaved by rival alliances, nor is there eagerness on the part of major European powers to push the expansionist agendas of Balkan clients. In fact, until recently both Russia and the principal members of the European Union repeatedly resisted calls to become involved militarily (beyond deploying small contingents as part of the UN peacekeeping forces in Croatia and Bosnia) in the Yugoslavian morass.

Any change in that situation is due to the Clinton administration's maladroit policy. Washington's obsession with bringing about a comprehensive peace settlement has created a dynamic in which not only the NATO powers but Russia as well have begun to play larger and more assertive roles in Bosnia. Russia is apparently expected to "deliver" the Serbs by pressuring them to make the requisite political and territorial concessions. U.S. leaders apparently assume that Moscow's activism will advance Washington's policy objectives—that the Russians will be cooperative junior partners in a U.S.-designed peace process.

But the administration is playing a dangerous game, since it creates the possibility of NATO and Russia eventually lining up on opposite sides in the Bosnian conflict.[32] Columnist Patrick Buchanan identifies the underlying risk, noting that Clinton's actions have "made the Balkan war what anyone with a sense of history hoped it would not again become: a playground for great power rivalry." The trend was worrisome, for "America has emerged as the patron of the Muslims, Moscow the champion of the Serbs."[33]

The adoption of a more activist, high-profile NATO policy toward Bosnia with the proclamation of the Sarajevo ultimatum in February 1994 placed Russia in an extremely uncomfortable position. Although the Yeltsin government had little desire to become entangled in a Balkan conflict, it concluded that NATO's increasingly coercive strategy could not go unchallenged without creating serious domestic political risks. The West's blatantly anti-Serb policy was resented by the Russian people as well as by the overwhelming majority of parliamentary deputies. Public opinion surveys taken in the spring of 1994 revealed widespread hostility to NATO's apparent course of action on the part of Russians across the political spectrum; the pro-Serb perspective was not confined to the ultranationalist fringe.[34] Perhaps more important, military leaders from Defense Minister Pavel Grachev on down were insisting that Yeltsin stand up to NATO and defend Russia's long-standing interests in the Balkans.

Faced with that situation, Yeltsin and his advisers tried to walk a foreign policy tightrope. They realized that, given the disarray of its armed forces, Russia was in no position to contemplate a military challenge to NATO's policies. Moreover, even mounting a direct political challenge—for example, by insisting on explicit UN Security Council authorizations before coercive measures could be taken against the Serbs and then, perhaps, vetoing some of those authorizations— could entail unacceptable costs. Among other consequences, such intransigence might jeopardize the delivery of promised financial aid from Western governments as well as approval of loans from the International Monetary Fund.

Precluded from thwarting NATO in more direct ways, Moscow opted to try a diplomatic ploy. Foreign Minister Andrei Kozyrev and other officials pressured the Serbs to accede to NATO's ultimatum and withdraw their tanks and heavy guns from the proclaimed exclusion zone around Sarajevo to avoid a dangerous showdown that would leave Russia little room in which to maneuver. At the same time, the Russians virtually demanded that the UN secretary general authorize the redeployment of several hundred Russian peacekeeping troops from Croatia to Sarajevo. That move caught the NATO powers completely off guard and placed them, in turn, in a politically delicate position. The United States and its allies had no desire to see Russian soldiers make an ostentatious entrance into Sarajevo as the saviors of the Serbs. At the same time, an effort to block what Yeltsin and Kozyrev insisted was a helpful gesture to dampen the Bosnian conflict would be evidence that NATO was trying to exclude Russian influence from the Balkans. Such a perception would cast a pall over the entire range of relations between the Western powers and Russia. Caught in that dilemma, NATO leaders publicly praised Moscow for its constructive action; privately, though, they were seething at what many of them considered Russian duplicity.

Russia's handling of the Sarajevo ultimatum was an example of astutely playing a very weak military, political, and diplomatic hand. Without provoking a confrontation with NATO—one that Moscow could not possibly win—the Yeltsin government nevertheless managed to reestablish Russia's position as a relevant factor in Balkan affairs. The Serbs seemed grateful (at least temporarily) that Moscow's intervention had saved them from NATO coercion; the mere fact that the Bosnian Serb forces only acquiesced to NATO's demands after that intercession appeared to demonstrate to the West that Russian influence would be vital if there was to be any hope of a peace settlement; and domestically the sight of Russian troops in Sarajevo quieted critics who had charged that the Yeltsin regime was a Western patsy. One of the worrisome consequences of Russia's maneuver, however, was the development that Buchanan identified: the great powers were beginning to line up on opposite sides in a Balkan conflict.

The increasingly difficult relationship between NATO and Russia became even more apparent a few weeks after the Sarajevo crisis when the alliance launched air strikes against Serb positions around Gorazde in April 1994. Yeltsin angrily denounced the action, especially because Russia had not been consulted. Following a conversation with Clinton, the Russian president informed reporters, "I insisted to Clinton time and again that such decisions cannot be taken without prior consultation between the United States and Russia."[35] Shortly thereafter, Moscow indicated that it would not sign the presentation documents approving Russian membership in the Partnership for Peace.

The primary motive for Yeltsin's condemnation was apparently not solicitude for the Serbs. Indeed, Yeltsin, Kozyrev, and other officials were becoming increasingly exasperated with what they regarded as intransigence and duplicity on the part of Bosnian Serb leaders. (That exasperation reached the boiling point later in April when the Serbs repeatedly violated promises to lift the siege of Gorazde. Their deceptions provoked Yeltsin's special envoy, Vitaly Churkin, to warn them not to treat Russia like a "banana republic.") But the NATO air raids once again had put the Yeltsin government in a precarious position. Although Russian officials seemed less than enthusiastic about their Serb "friends," Moscow could not abandon them to NATO's tender mercies without serious domestic political repercussions. Equally important, a failure to react to what seemed to be a calculated NATO snub would have eroded much of the diplomatic

105

prestige accumulated during the Sarajevo episode. Consequently, Russia again adopted an assertive position. The incident also afforded Yeltsin a pretext for postponing action on joining the Partnership for Peace—which had generated unexpectedly intense domestic controversy when the decision to sign had been announced a few weeks earlier.

Yeltsin's ability to continue walking the diplomatic tightrope in the Balkans was tested again within days of the initial air strikes around Gorazde. NATO leaders indicated that the alliance was prepared to extend Sarajevo-style exclusion zones to protect Gorazde and other Muslim safe havens and hinted that such action might occur whether Russia approved or not. As in the case of the Sarajevo incident in February, there was little Russia could do unless it was prepared to provoke a confrontation it could not possibly win. As a face-saving response, Moscow stated that it had been duly consulted by the NATO powers and, therefore, endorsed the exclusion zone plan despite some misgivings that it might intensify rather than dampen the Bosnian conflict. To rinse the taste of bile from that concession, the Western allies endorsed, in principle, Yeltsin's proposal to create a formal "contact group" consisting of American, British, French, German, and Russian officials to draft a new peace plan for Bosnia and press the feuding parties to arrange a viable cease-fire.[36] Once again, Moscow had managed to save face when it had little political or military leverage.[37]

There were indications, however, that domestic support for the strategy of acquiescing to NATO's threats of military coercion of the Serbs in exchange for a higher profile Russian diplomatic role was thin, at best. Although Kozyrev and Churkin indicated that Russia would not object to air strikes if the Serbs violated the exclusion zones, Grachev openly condemned any further coercive measures, and the parliamentary opposition was incensed.[38] Nationalist critics predictably condemned the Yeltsin government for a craven capitulation to NATO's demands.

Despite the avoidance of an open breach between NATO and Russia over the Gorazde incident, there is growing evidence of tension. Russia's long-standing cultural, religious, and political ties to the Serbs, combined with the pressure Yeltsin is receiving from domestic critics, mean that Moscow cannot easily endorse NATO's wholesale coercion of the Bosnian Serbs—much less attacks on Serbia itself. Persisting in a strategy of military escalation, therefore, will at the very least risk damaging relations with Russia, and perhaps fatally undermining the Yeltsin government. Indicative of that government's continuing effort to walk the diplomatic tightrope has been its role in the new contact group. When the Bosnian Serbs rejected the group's latest peace proposal in July 1994, Moscow nevertheless blocked efforts (principally by the United States and Germany) to respond with military measures. Instead, Russian leaders insisted on a far milder response—a tightening of economic sanctions.[39]

At worst, a confrontational policy on the part of NATO could begin to replicate the conditions that led to the conflagration in 1914. Although Russia's weakened military condition makes that scenario unlikely in the short term, it is testimony to the myopia of NATO, especially U.S., policy that the risk has been created at all.

NATO's muddled intervention in Bosnia, particularly the incremental escalation of that mission, constitutes a compelling argument against the expansion of the alliance into Eastern Europe. It should serve especially as a warning to people such as Sen. McCain, who has presented trenchant criticisms of military action in Bosnia but who paradoxically supports the enlargement of NATO. Numerous Bosnia-style quagmires are precisely what the alliance can expect if it ventures into Eastern Europe.

Attempting to pacify that region will inevitably involve the NATO powers in conflicts fraught with moral ambiguity—in which none of the parties wear white hats or black hats, but all hats are a dirty shade of grey. The alliance will encounter intractable struggles, with deep and tangled historical roots—conflicts in which the feuding factions are willing to fight on year after bloody year and seem impervious to reason or to the notion of compromise. NATO will be called upon to use military force in pursuit of the most nebulous political objectives.[40] And if the alliance adopts an interventionist policy, it will risk a collision with Moscow as apprehensive Russian leaders react to evidence that the Western powers are seeking to displace Russia from a region where it has long-standing political, religious, and security interests. All of those problems are already apparent in NATO's Bosnia adventure. The costs and risks of such missions throughout Eastern Europe would vastly exceed any likely benefit.

# 6. The Politics of Preservation

The vigorous campaign by NATO partisans to expand the alliance into Eastern Europe is only the latest manifestation of a search for alternative missions. The sudden collapse of the Soviet empire, followed rapidly by the disintegration of the USSR itself, deprived NATO of its raison d'être and caught its supporters off guard. Nevertheless, they soon reacted in the way one would expect defenders of a threatened vested interest to respond. In a practical demonstration of the public choice economic theories that won James Buchanan the 1986 Nobel Prize in economics, they circled the intellectual wagons to protect their beleaguered institution.

Virtually without exception, their operating premise was that NATO must be preserved despite the political upheaval that had transformed beyond recognition not only the European geostrategic landscape but the entire international system. That is an assumption that has not wavered with the passage of time. Brookings Institution scholar Catherine McArdle Kelleher expresses that premise, warning that most of the changes thus far adopted by NATO "do not go to the heart of the post–cold war alliance dilemma—that is, how to define a new over-arching purpose that will garner public support and that goes beyond preparing for defense against uncertainty and then transforms the alliance accordingly."[1] Kelleher's reasoning is typical of that of ardent Atlanticists; they do not even entertain the possibility that NATO may no longer serve American interests—much less that the alliance may not be relevant at all—in the post–Cold War era. Rather, the issue is seen primarily as a marketing problem—coming up with a new mission that can "garner public support," that is, be sold to the public.

## Early "New Mission" Prototypes

Some of the initial new rationales for NATO testified to the desperation as well as the creativity of Atlanticists. Former assistant secretary of state Robert D. Hormats, for example, suggested that Western leaders "expand the range of issues on which NATO engages the common efforts of the European and North American democracies—from student exchanges, to fighting the drug trade, to resisting terrorism, to countering threats to the environment."[2] Former U.S. ambassador to NATO David Abshire also insisted that NATO could coordinate the transfer of environmental-control and energy-conservation technology to the East, thereby benefiting the global ecology."[3]

Even some NATO preservationists seemed uncomfortable about stressing such nonmilitary roles as the primary reason for continuing what was supposedly a military alliance. Gen. Daniel Graham, the principal promoter of the Reagan administration's Strategic Defense Initiative, suggested a more plausible alternative mission: preventing unstable or aggressive Third World states from acquiring ballistic missiles and weapons of mass destruction.[4] Graham was not entirely clear about how NATO would carry out such a mission, whether through tighter controls on the export of militarily useful technology, the imposition of economic sanctions, or a more drastic version of coercive nonproliferation—the launching of preemptive military strikes. Nor was he clear why NATO was the appropriate institutional mechanism for dealing with such problems. Nevertheless, he was at least articulating a new security mission for the alliance.

The onset of the Persian Gulf War raised another possible new role. Pro-NATO spokesmen contended that the alliance was quite useful to the U.S.-led military effort against Saddam Hussein. At the very least, they insisted, American troops and equipment stationed in Europe as part of Washington's NATO commitment would be closer to the scene for deployment in future Middle East crises. Some even held out the more ambitious prospect that the alliance might become an institutional arrangement for coordinating a Western response to breaches of the peace in that volatile region. Former president Richard M. Nixon became an outspoken advocate of that course.

> [NATO's] creators did not envision that by specifying that the NATO commitment applied to Europe and North America, the alliance would operate only within a strict boundary. . . . Today, as demonstrated in the Persian Gulf, challenges to Western interests can arise half a world away. If NATO adheres mindlessly to artificial geographical restrictions, we will simply be shooting ourselves in the foot, compromising our interests to

legalism. . . . While European defense must remain NATO's
core mission, so-called "out of area" security cooperation must
become its cutting edge.[5]

Such arguments foundered on the rocks of some troublesome
realities. For one thing, France and other key member states had
prevented NATO from embracing the gulf intervention as an official
alliance military operation—insisting that NATO's defense mandate
applied only to the North Atlantic region as defined in the treaty.
Consequently, those NATO members that contributed to the anti-Iraq
coalition effort did so on an individual, national basis. That did not
bode well for the alliance's taking on other Middle East military
crusades in the future.

The manifest reluctance of Washington's NATO partners to give the
alliance security responsibilities outside Europe raised questions even
about the alleged advantage of stationing U.S. troops on the Continent
for possible deployment in a crisis elsewhere. It was not certain that a
European host government would be receptive if U.S. forces stationed
in its country, ostensibly for the defense of Western Europe, were
dispatched to the Middle East or some other region—perhaps even to
implement a policy that the host government opposed. It was indicative of the problems associated with the entire issue that U.S. and other
NATO leaders rarely invoked the out-of-area justification for a continuing U.S. troop presence when they addressed European audiences.
They apparently understood that there was scant public support in
Western Europe for using the Continent as a staging area for U.S. forces
whose mission was other than the defense of Europe.

**The Quest for Survival**

The implausibility or divisive nature of the other candidates for
NATO's new mission made the option of expansion into Eastern
Europe the winner almost by default. Although there are serious
questions about the feasibility of that mission as well—and the existing
NATO members are by no means united in undertaking the requisite
burdens—it is the most credible suggestion for perpetuating an
otherwise moribund alliance.

Secretary of State Warren Christopher hinted at the institutional
survival motive underlying that alternative mission as well as previous
versions in an address to NATO foreign ministers in December 1993.
Outlining the features of the Partnership for Peace, he warned that the

alliance must make a historic choice. "That choice is whether to embrace innovation or risk irrelevance."[6] Sen. Richard Lugar (R-Ind.) and other NATO partisans have stated flatly that the alliance must go out of area or go out of business. The search for missions and the "preservation at all costs" mentality within NATO are also illustrated by the comment of one high-ranking alliance official. He mused that in assessing the various security issues facing post–Cold War Europe, "We're working on the problem—even though we're often not quite sure what it is."[7]

The West European states have fairly clear motives for wanting to preserve NATO. It is the institutional mechanism for continuing a lucrative U.S. subsidy of their defense—a subsidy that spares them tens of billions of dollars in additional military expenditures that they might otherwise have to approve to adequately protect their security. The alliance and the American military tie it symbolizes are a long-term insurance policy for Western Europe. Although there is currently no serious threat to the security of the region, NATO is an assurance of U.S. assistance in the event such a threat ever reemerges.

The enthusiasm of the Central and East European countries for NATO is equally explicable in terms of national interest. Not only would membership in the alliance afford them protection against larger and more powerful neighbors, but many East Europeans see NATO membership as a way of entering West European institutions—especially the European Union's economic structures—through the back door.

Support within the American political and foreign policy communities for preserving—much less expanding—NATO is more difficult to fathom if one focuses only on a national interest criterion. At the least, one would expect a vigorous, wide-ranging debate about whether the alliance serves American interests in the post–Cold War period and whether the benefits of continued alliance leadership outweigh the costs and risks. The virtual absence of such a debate despite the collapse of the Soviet enemy is striking. For an explanation, one must look beyond national interest considerations and examine the intellectual and institutional motives of a network of entrenched interests.

Part of the determination to preserve NATO at all costs is a manifestation of a larger identity problem afflicting members of the American foreign policy elite. As Leslie H. Gelb observes, foreign policy professionals have not adjusted well to the post–Cold War era.

"The foreign policy elite—children of the American empire, a generation reared in the excitement and primacy of foreign policy—feel both disinherited by the Cold War's end and genuinely alarmed about a world drifting toward perpetual violence."[8] Their primary response has been to cling tenaciously to familiar policies and institutions.

The politics of preservation is not confined to NATO. Indeed, the campaign to extend the life expectancy of the alliance is merely one component—although an especially important one—of a larger effort by the U.S. national security bureaucracy and its network of supporters and beneficiaries to preserve as much as possible of Washington's global activist role. There are many other examples of the preservationist and "mission procurement" mentality.

The Central Intelligence Agency and other components of the U.S. intelligence apparatus have not only tenaciously fended off attempts at budget reductions, they insist that their functions are even more important in the post–Cold War world than they were during the struggle with the Soviet Union. According to CIA director R. James Woolsey, although the free-world nations have slain a ferocious dragon, that menace has been replaced by "a bewildering variety" of poisonous snakes.[9] At various times, the CIA and its supporters have identified terrorism, drug trafficking, nuclear weapons proliferation, Islamic fundamentalism, and the reliable catchalls of instability and uncertainty as threats that justify retaining the intelligence bureaucracy's Cold War levels of spending and personnel.

Woolsey contends that major cuts have already taken place, pointing out that the number of intelligence personnel is scheduled to shrink by 22.5 percent by the end of the decade, and argues that further reductions would imperil national security. When pressed, however, the director concedes that the cuts will merely bring the staff back to the size it was before the massive buildup during the Reagan administration.[10] In other words, the number of CIA personnel will return to "normal" Cold War levels.

The agency has also ranged far afield from traditional (albeit vague) security problems in its search for alternative missions. It has at least flirted with adopting economic espionage as a new raison d'être. Robert Gates, CIA director during the final years of the Bush administration, seemed somewhat hesitant to pursue that mission. Gates feared that it could damage the agency's relations with friendly governments, since most economic espionage would be directed at

113

other democratic capitalist countries. He was also concerned about the political problems of sharing intelligence data with American businesses. Providing information to one firm without giving the same information to all of its competitors would inevitably expose the agency to charges of favoritism and involve it in an assortment of nasty controversies with aggrieved members of Congress.[11]

Under Woolsey's leadership, however, the CIA appears to have become less timid about entering the arena of economic espionage. Woolsey himself denies that the agency is moving in that direction, except for analyzing such macroeconomic issues as currency flows. But there are indications that the CIA may already be more deeply involved in various forms of economic intelligence than either the public or Congress has been led to believe.[12]

The mission procurement approach was evident as well in the Pentagon's celebrated "Bottom-Up Review" of strategy and force structure conducted in 1993. In a memo to Secretary of Defense Les Aspin, Under Secretary of Defense Frank Wisner assured his boss that the review would "use the analytical expertise in OSD [office of the secretary of defense] and the Joint Staff to . . . provide more thorough and compelling descriptions of the New Dangers."[13]

In other words, what purported to be a dispassionate, bottom-up assessment of the global threat environment was largely an exercise that hyped "new dangers" to justify the Pentagon's predetermined force and budget proposals. Not surprisingly, the "Bottom-Up Review" reached the conclusion that the world was an extraordinarily dangerous place and that the United States had to maintain forces capable of waging two major regional wars almost simultaneously. The projected budgets for such a force during the remainder of the 1990s would keep U.S. military spending at 85 percent of the average during the Cold War. Wisner's approach was reminiscent of the cynical comments of Joint Chiefs of Staff chairman Colin Powell two years earlier. In an unusually candid moment, Powell stated: "I'm running out of demons. I'm running out of villains. I'm down to Castro and Kim Il Sung."[14]

Although Powell may have had a depleted stock of international demons with which to frighten the American people, Wisner and his counterparts have been able to come up with an abundance of new dangers to serve a similar purpose. For the NATO preservationists, the principal new dangers are those of Russian neo-imperialism (symbolized by the rantings of "Mad Vlad" Zhirinovsky) and the vague but menacing specter of European "instability." Both are held to justify not only preserving the existing version of NATO indefinitely but expanding the alliance and giving it the mission of preventing conflict in Eastern Europe.

In some respects, the instability mission is better than the Cold War objective of containing the Soviet Union for ensuring NATO's permanence. Given the abundance of unresolved grievances, territorial disputes, and ethnic and religious quarrels throughout Eastern Europe, there is no danger that such a mission will suddenly disappear, as the Cold War did. It will be there decade after decade, justifying high levels of military spending, swollen personnel rosters for various components of the national security bureaucracy, and a bottomless reservoir of problems to be addressed by foreign policy practitioners.

**Public Choice Factors**

The search for alternative missions for NATO is both the intellectual and the institutional response of vested interests that feel threatened by the end of the Cold War. Their reactions have thus far been textbook examples of the processes described in public choice economic theories. Nobel laureate James M. Buchanan, his long-time colleague Gordon Tullock, and other public choice economists have demonstrated that the activities of any interest group are "a direct function of the 'profits' " the group expects "from the political process."[15]

Contrary to the contentions of some political scientists who posit a benevolent government populated by public servants who base their decisions solely on what is in the national interest, Buchanan and his colleagues argue that government bureaucracies and their constituents react to incentives much as participants in the private sector do. Only the nature of the incentives is different. Since there is no profit motive or "bottom line" to use as a reference point for judging the efficacy of specific policies or actions, government agencies have different criteria for measuring success. The principal measurements are budgets and staff levels. High (and expanding) budgets and increases in staff are considered signs of agency health.

There is also a public-sector analog for the marketing strategies pursued by for-profit businesses. A private-sector enterprise typically seeks to penetrate new markets or gain a greater share of existing markets. In a similar fashion, bureaucracies seek out new missions or identify new problems in existing areas of responsibility. Again, the principal measure of success is a growth of the number of important issues that the agency "manages."

Gradually, policy networks or "iron triangles" grow up around major issue areas. Those triangles—three-way interactions involving elements of the permanent bureaucracy, members of Congress, and special interest lobbies—are dedicated to promoting larger agency budgets and missions and to distributing concentrated benefits to members of the network. Those benefits can be either tangible or intangible. In the case of the national security policy triangle, an example of the former would be lucrative contracts for defense firms and the resulting political advantages to senators or representatives who can take credit for bringing home such contracts to their states or districts. An example of the latter would be the heightened prestige and inflated sense of importance of leaders of the defense and foreign policy bureaucracies as well as "friendly" consulting firms, think tanks, and key members of the news media.

The nongovernmental members of the policy triangle become crucial transmission belts between the government and the general public. They form the core of issue networks, groups that take an intense interest in some aspect of public policy. Issue networks help policy triangles disseminate ideologically charged information that legitimizes policies and enables the triangles to dominate the public debate on particular topics.[16] The fact that issue networks are not overtly affiliated with the government creates an aura of independence that gives their studies, news stories, and speeches a credibility that official government pronouncements might lack.

The cozy relationship between the policy triangles and their allied issue networks is intensified by the revolving door phenomenon. Prominent journalists, academics, and think tank experts form the reservoir of personnel for policymaking positions in the government. After a period of time in "government service," they typically return to positions in the private-sector component of the issue network, where they comment on the wisdom of policies being pursued by the colleagues they left behind or the new team of policymakers. The incentives inherent in the revolving door system ensure that any inclination to make harsh assessments will be tempered by the fact that the critic may wish to return to government service someday.

It would be an overstatement, of course, to suggest that all members of an issue network are in agreement on all issues. Especially within large and diffuse issue networks, there can be noticeable—sometimes even bitter—factionalism. That is particularly true if the political elite itself is badly split on policy options (as during the final years of the Vietnam War) or if members have competing interests in a specific issue. Nevertheless, even though there may sometimes be divisions or shifting coalitions within an issue network, the various factions still have a common interest in preserving and expanding the issue area in which they operate. Any development that makes the issue seem less relevant threatens the interests of all participants. The demise of consulting firms, research institutes, and publications (e.g., *Global Affairs*) that were too narrowly focused on the security threat posed by the Soviet Union is a prime example of that danger.

Shared interests tend to keep factionalism within bounds. Moreover, issue networks are quintessential examples of the "good old boy" phenomenon. Those policy experts who are ambitious for career advancement, especially obtaining prestigious appointed positions in government, learn early that unorthodox views and anything other than the mildest criticisms of the conventional wisdom are not the path to advancement. The revolving door intensifies that tendency and minimizes not only factionalism but original thinking.

Nowhere are those characteristics more evident than in the national security bureaucracy and its affiliated issue network. The national security function of the U.S. government during World War II and the Cold War produced one of the most powerful and tenacious policy triangles and issue networks. At the center is the military budget, which reached $300 billion annually during the final phase of the Cold War and is still more than $260 billion. (Lesser but still significant components of the issue area are the general foreign affairs budget and the foreign aid budget.) Not surprisingly, a wide-ranging and influential constituency developed to promote the national security agenda. As in most other cases, the motives of the direct and indirect beneficiaries of large defense and foreign policy budgets have been a mixture of genuine concern—it is after all vital to protect the security of the American people from foreign enemies—and careerist and other self-serving objectives.

The one thing that the national security policy triangle and the bulk of the associated network have in common is a determination to see the levels of U.S. military spending reached during the Cold War preserved as much as feasible in the post–Cold War period. Keeping NATO intact, and finding credible new missions for the alliance, is crucial to preventing serious erosion of the military budget. Washington's commitment to defend its European allies costs approximately $90 billion a year. That cost includes not only the expense of maintaining the U.S. military contingents stationed in the European theater but also the expense of the combat forces earmarked as reinforcements in the event of a major European war, as well as the airlift and sealift units needed to transport them.

Without the NATO commitment, the justification for more than one-third of the U.S. military budget would disappear. Advocates of high levels of military spending would undoubtedly argue that the United States still needs all of the military forces currently contemplated in the budget projections, but many Americans would likely be skeptical. Public opinion surveys already show widespread support for going beyond the modest spending reductions proposed by the Clinton administration and making more substantial cuts. Even the Pentagon's boosters concede that it is increasingly difficult to muster public support for large defense budgets in the absence of a clearly identifiable threat to America's security. Invoking the vague specter of global instability simply does not have the same impact as pointing to a large tangible menace such as the Soviet Union and its communist allies.

Some portions of the public—and even a few former officials and members of Congress—are beginning to question the wisdom of maintaining oversized military budgets. They ask why the United States spends more than $260 billion on the military while Japan and Germany, the world's number two and three economic powers, spend merely $39 billion and $31 billion, respectively. As Lawrence Korb, a former assistant secretary of defense in the Reagan administration, points out, the United States spends more on the military than the rest of the industrialized world combined.[17]

It would be exceedingly difficult to justify such a disparity without the image of a U.S.-led NATO dealing with dangerous crises in Europe and preventing those crises from leading to a rerun of the two world wars. For that image to be credible, NATO needs a security mission that is more relevant than keeping armored divisions in Germany to deter an invasion from a defunct Warsaw Pact led by a Soviet Union that no longer exists. That is the operational significance of warnings by Lugar and others that NATO must go out of area or go out of business. Lugar himself is surprisingly direct about the domestic political motives underlying the call for new missions. "A credible American commitment focused on territorial defense against a nonexistent threat," Lugar warns, "cannot long be politically sustained. . . . If only for domestic political reasons, a new rationale," one revolving around new missions, "may be essential to halting the erosion of support for NATO."[18]

By expanding its jurisdiction and security role into Eastern Europe, the alliance would acquire an abundance of frightening crises that would have to be "managed." The prospect or reality of American troops being at risk in such crises could then be used to justify an oversized U.S. military budget in perpetuity.

Washington's NATO commitment is the linchpin that secures future benefits for the national security policy network. Without the alliance, the entire U.S. military budget would be open to unprecedented scrutiny and controversy. Among other things, it would prove increasingly difficult to sustain adequate public support for other U.S.-led alliances. Americans would predictably ask why, if the United States was no longer subsidizing Europe's defense, it should continue to subsidize the defense of such prosperous East Asian allies as Japan and South Korea. It would be hard for even the most creative Pentagon supporters to make a plausible case for stationing 100,000 troops in East Asia if the United States did not have a comparable force in Europe—the region long cited as the most important to America's own security and well-being. For obvious reasons, the beneficiaries of the national security policy triangle and most members of the issue network do not want a debate on such matters.

### The Tyranny of Wilsonian Status Quo Thinking

Although career and financial factors are crucial in explaining the intensity of the determination of America's NATO partisans to protect and expand their institution, more subtle ideological, intellectual, and perhaps even psychological considerations also play an important role. The U.S. foreign policy community grew accustomed to the United States' being the leader of the free world throughout the Cold War. The end of that conflict has placed the United States at a crossroads in terms of its political-military role. One possibility is to exploit America's status as the world's sole remaining superpower and adopt an even more activist policy. The disintegration of the only country that could plausibly mount a serious military challenge to U.S. preeminence makes such a course seem feasible. The other possibility is to adopt a less interventionist approach, cast off the excessive security burdens acquired during the Cold War, enjoy the financial windfall of a "peace dividend," and adopt a foreign policy that is more typical of an ordinary great power in a multipolar international system.

Most members of the American foreign policy community have endorsed the first option. Indeed, they tend not only to reject the second option but to regard it as personally threatening. To them, America's continuing global leadership is not merely a matter of national pride—although there is often a good measure of that. They also view it as a reflection of their own importance. American policymakers and nongovernmental experts have long been treated with the respect, even deference, due representatives (official or unofficial) of a superpower. A more circumscribed role for the United States threatens their special status. That sense of potential loss may be one reason why prominent foreign policy professionals react with uncharacteristic emotionalism both to suggestions by iconoclastic American scholars that U.S. power is in relative decline and to any manifestations of greater independence or assertiveness on the part of Washington's Cold War era allies.

A more restrained political and military role for the United States also runs counter to several decades of immersion in the alleged virtues of "enlightened internationalism," especially its crusading Wilsonian variant. Most members of the foreign policy community have come to regard as axiomatic the notion that the United States cannot really be secure unless the international system is stable and peaceful. Even such figures as Henry Kissinger and Zbigniew Brzezinski, who regard themselves as "realists," subscribe to the proposition to a surprising degree.[19]

There is, however, scant evidence to support the "seamless web" theory of U.S. security. Throughout most of its history, the United States had to operate in an international system in which the overwhelming majority of nations were undemocratic and conflicts (often pursued for the most cynical of motives) were the norm. Even today, democracy is practiced in only a minority of countries, despite the strides made following the collapse of the Soviet empire. And the number of conflicts has spiraled. There are some 48 armed conflicts in progress, including 29 "major" wars—those producing at least 1,000 fatalities. Yet most of those struggles have had little or no adverse effect on America's security.[20]

Wilsonian internationalists rarely pause to question their assumption that U.S. security requires a high degree of international stability, much less whether such an objective can be attained at an acceptable cost. (In that regard, it is hardly comforting that National Security Adviser Anthony Lake has described the Clinton administration's approach to world affairs as a "pragmatic neo-Wilsonian" policy.)[21] Despite its dubious assumptions and serious conceptual defects, Wilsonianism remains the essence of the conventional wisdom on foreign policy in the United States.

The assumption that global instability per se poses a threat to U.S. security inevitably inclines foreign policy practitioners to favor not only a highly activist U.S. military role but also the preservation of alliances and other arrangements (such as the United Nations) that have the objective of preserving stability. NATO, long the most prominent of Washington's alliances, is considered almost sacrosanct.

In addition to the policy rationales that NATO's supporters routinely cite to justify perpetuating the alliance, there is a complex array of underlying factors. Those include mundane careerist and institutional objectives, manifestations of an identity crisis in the American foreign policy community, and the tenacious grip of Wilsonian myths about international politics. Together they create a powerful force to preserve and extend the alliance, despite the probability of undesirable consequences for the American people.

# 7. Beyond NATO: Encouraging "Europeans-Only" Successors

Former British diplomat Jonathan G. Clarke makes the perceptive observation that "a range of geostrategic, political, regional, and economic considerations suggests that NATO does not provide the best answer to contemporary and future American security needs in Europe. Nor does it guarantee continuing American influence in Europe." Clarke concludes, "If NATO did not already exist, it is doubtful that Washington would now invent it."[1]

NATO enthusiasts, of course, see matters differently, contending that there is no credible substitute for the transatlantic alliance. To them NATO is, quite literally, irreplaceable, But Clarke and other critics make a strong case that other existing security organizations would be more than adequate substitutes.

One possible successor might be a strengthened version of the Conference on Security and Cooperation in Europe (CSCE). The framework of that organization was established in July 1973 by the foreign ministers of 35 nations (the United States, Canada, and all European countries except Albania). Two years later, with the adoption of the Helsinki Final Act by the heads of state of those nations, the CSCE was given a tripartite thematic mission: the promotion of security, economic cooperation, and human rights. With the collapse of the Warsaw Pact and then the Soviet Union itself, the membership of the CSCE expanded to its present roster of 56 members—including several of the former Soviet republics that are technically in Asia rather than Europe. The end of the Cold War division of Europe into hostile military blocs has also led to proposals for making the CSCE an effective pan-European collective security organization.

One might be justifiably skeptical about the CSCE's becoming a credible successor to NATO. It exhibits many of the defects that an enlarged NATO would, most notably an excessively diverse membership and an overextended geographic coverage. The current agreement also requires all decisions, even on relatively minor matters, to be reached by unanimous consent, which would be an inherent obstacle to effective action in a crisis.[2] The initial record of the organization, most notably the failures to prevent or resolve the civil war in the former Yugoslavia and the conflict between Armenia and Azerbaijan, does not inspire a great deal of confidence either. Nevertheless, even the CSCE may have some utility as a forum for the airing of political grievances, as a mediation and conciliation service, and perhaps for limited peacekeeping operations to police settlements to conflicts that have been reached by the belligerents. (If the United States remains a member, however, it must emphasize that it will not participate in military sanctions or CSCE peacekeeping operations.)

### NATO's Possible Successor: The WEU

An association that has considerably more potential as a practical security organization is the Western European Union (WEU), a military alliance of nine West European states. The roots of the WEU lie in the five-power Brussels Treaty of 1948, which linked Britain, France, and the small Benelux countries in a mutual defense pact. That treaty represented the first serious attempt by the West European powers to come to grips with their difficult post–World War II security situation. The alliance was directed against two potential adversaries. The Soviet Union represented the more immediate threat in the minds of most West European leaders, since Moscow had already subjugated Eastern Europe, maintained military forces far more numerous than were arguably needed for defensive purposes, and had developed close ties with powerful fifth-column communist movements throughout Western Europe. Some leaders in the Brussels Pact countries, however, also worried that a revanchist Germany might pose a serious security threat in the long term. The Brussels Pact pooled the strength of the West European states to meet both potential dangers.

The deepening Cold War quickly made concerns about Soviet intentions the dominant factor. If Western, especially British and American, policymakers had not stressed the need for a powerful transatlantic alliance to counter the Soviet threat, the Brussels treaty organization might well have become the dominant institutional arrangement for protecting the security interests of Western Europe. NATO, however, rapidly eclipsed the Brussels Pact in importance. The most significant role played by the Brussels Pact was as a convenient vehicle for incorporating West Germany into the West's security system in 1955 and thereby permitting West German rearmament. With the inclusion of West Germany, the member states renamed the organization the Western European Union.

In most other respects, the WEU remained virtually dormant until the mid 1980s. To the extent that it had a meaningful function at all, it acted mainly as an institutional bridge between the members of the European Community and Britain, which for much of that period was not a member of the EC. Representatives on the WEU's governing Assembly were usually back-bench politicians who had little or no influence at home. So moribund had the organization become that the more significant Council of Ministers (consisting of the foreign and defense ministers of each member) did not even meet in formal session between 1973 and 1984.

The renaissance of the WEU began at a meeting of the Council of Ministers in Rome in October 1984. That meeting occurred in the context of two developments that were both worrisome and annoying to the West European countries. One was the sharp deterioration in East-West relations that followed the Soviet invasion of Afghanistan and the election of Ronald Reagan as president of the United States. The other was a growing sense that the interests of Western Europe and the United States did not always coincide, that Washington was all too willing to run roughshod over the wishes of its allies when such differences emerged, and that because of their lack of political unity the Western Europeans could do very little about such cavalier treatment.

During the early and mid-1980s the Soviet Union and the United States seemed to be engaged in the most acrimonious confrontation since the initial decade of the Cold War. Large portions of European public opinion and not a few West European officials became increasingly apprehensive that a war between the two superpowers was no longer out of the question. Although in retrospect it is evident that such fears were exaggerated, they seemed valid to many Europeans at the time.

Concern about the tense state of U.S.-Soviet relations also led to a greater awareness of Western Europe's acute security dependence on the United States. Most European leaders worried not only that Washington might engage in reckless action that could embroil its European allies against their will in war with the Soviet Union— although that was certainly a factor. The concern was both deeper and more subtle. At root was a pervasive sense of military, and even

125

political, impotence—the belief that the United States was not overly concerned about West European views on important security issues, and that Western Europe had no institutional mechanism with which to coordinate its own policies and display a common front to Washington if the Europeans disagreed with U.S. policy.

Washington's dominant role in NATO accentuated Western Europe's dependent status. Historian Ronald Steel captures that reality perfectly, noting that NATO had always been "a kind of Lincoln's Cabinet: every member would solemnly express its opinion, but the American president had the only vote that mattered."[3] By the mid-1980s France and several other West European nations were no longer content with that arrangement.

Friction in Washington's relations with its NATO allies transcended the usual high-profile security issues. For example, the Reagan administration strongly opposed the construction of a natural gas pipeline linking the Soviet Union and Western Europe, fearing that it would make the democratic nations of Europe dangerously dependent on Moscow for their energy supplies. Although the United States viewed the pipeline project as a NATO security issue, the West Europeans saw it primarily in economic terms. To them the pipeline offered a cheap and abundant source of energy, which would lower the production costs of major industries and make European firms more competitive in the global marketplace. Some West Europeans began to wonder if the U.S. concern about the alleged potential of the pipeline to make their countries vulnerable to Soviet blackmail did not conceal more mundane motives, such as a desire to protect the market shares of American companies. In the end, the allies brushed aside Washington's objections—an unprecedented act of defiance that was barely papered over by joint communiqués with the United States.

The perceived need to create a distinctly West European security identity was a major factor in the 1984 decision to revitalize the WEU. A more direct solution would have been for the Europeans to simply have given security responsibilities to the European Community, as the French favored, but objections from Britain and other members who wanted the EC to remain focused solely on economic and political issues made that step impossible. Giving the WEU a modestly greater security mission was deemed an acceptable compromise.

The initial steps taken at the 1984 meeting were relatively tame. A "Statement of Rome" issued by the ministers outlined the tasks of the revived WEU: an assessment of the Soviet threat, increased European arms collaboration—including efforts to standardize weapons and training—and the formulation of European views on arms control issues and East-West dialogue. Although such steps were hardly a serious challenge to NATO's preeminence in Western Europe's security affairs, they did hint that the Europeans were beginning to move toward greater independence and that the United States could no longer routinely treat its allies as protectorates that sometimes needed to be humored but rarely taken seriously.

Further evidence that the Western Europeans were beginning to chafe at Washington's overbearing attitude came in 1987 with a second round of efforts to reinvigorate the WEU. It was indicative of the changing climate of European opinion that the proposal to further strengthen the WEU came from the most unlikely source—Great Britain, which had previously stressed NATO's primacy as well as its own special security relationship to the United States. Other signs of greater West European assertiveness were not long in coming. In August Britain and France persuaded several WEU members to send naval forces to the Persian Gulf to assist in the U.S.-led escort of oil tankers. Although Washington welcomed such support for its gulf policy, U.S. officials should have noticed that the action was taken under WEU auspices rather than under NATO's banner or by individual national actors.

Two months later the WEU's Council of Ministers approved a strongly worded "Platform on European Security Interests." Although that document still proclaimed the WEU's institutional subordination to NATO, it also contained provisions that aimed at a stronger, distinctly European, role in security matters. The platform stressed the need for the retention of some nuclear forces and specifically noted the growing importance of the independent British and French nuclear arsenals. It also advocated an increase in conventional forces in Western Europe to maintain a credible deterrent against the Soviet Union and its Warsaw Pact allies.[4] The following year the WEU increased its roster of members for the first time since the mid-1950s, admitting Portugal and Spain. Significantly, Spain was permitted to join the WEU without joining NATO's military command—an action that underscored the WEU's growing independence in operational matters.

Ironically, whereas the poor state of U.S.-Soviet relations in the early 1980s had provided an important impetus for the initial steps to revitalize the WEU, the rapidly warming relations between the superpowers after the emergence of Mikhail Gorbachev as the new Soviet leader was the impetus for the subsequent, more important steps. Fear of a U.S.-Soviet confrontation that would disregard the wishes and interests of Western Europe had been replaced by an equally potent fear of U.S.-Soviet condominium that would likewise ignore West European interests. The sense of dependence and impotence was the unifying theme to what otherwise seemed a schizoid European response to the behavior of the superpowers.

Two developments intensified Western Europe's uneasiness. One was the 1986 Reagan-Gorbachev summit in Reykjavik, Iceland, which developed an extraordinary enthusiasm on the part of both participants for radical disarmament schemes. At one point the two leaders seriously discussed eliminating all strategic nuclear weapons in their countries' arsenals (the so-called double-zero option). An agreement in principle on such a revolutionary change foundered only because Reagan refused to abandon his Strategic Defense Initiative, the plan to develop a comprehensive shield against ballistic missiles.

The double-zero option discussions, commenced without Washington's engaging in prior consultations with its NATO allies, sent shock waves throughout West European capitals. The Europeans viewed the shield provided by the U.S. strategic arsenal as crucial to the security of their region. The only alternative to the protection afforded by such extended deterrence appeared to be a massive build-up of West European conventional forces combined with a major expansion of the British and French nuclear arsenals (and perhaps the development of nuclear weapons by other European powers as well). From the perspective of the West Europeans, Reykjavik signaled a U.S. desire to decouple its security from that of Western Europe and reduce America's exposure to the nuclear risks entailed in its commitments to defend the NATO allies. Even worse, the United States seemed prepared to undercut its allies' security position without warning, much less candid discussion.

That impression was reinforced by the conclusion of the Intermediate Nuclear Forces (INF) treaty between the United States and the Soviet Union. The INF agreement provided for the removal of all U.S. intermediate-range missiles from Western Europe in exchange for the withdrawal of similar Soviet weapons from Eastern Europe. The accord was most distressing to the West European governments, who regarded the U.S. intermediate-range missiles as a guarantee that any conflict in Europe would immediately escalate to the nuclear level. According to the logic of deterrence theory, the certainty that a Soviet invasion of Western Europe would go nuclear—thereby triggering the use of the U.S. strategic arsenal—was the best guarantee that such an attack would never take place. Any development, such as the INF treaty, that eliminated the tangible linkage to the U.S. strategic arsenal undermined that certainty and threatened to decouple U.S. and West European security interests. Without that linkage, there was a possibility that the United States would renege on its NATO obligations in a crisis or would at least try to fight a limited war confined to the European continent—something that the West Europeans understandably regarded with little enthusiasm. As had the Reykjavik negotiations, the INF treaty suggested that the United States sought to reduce its own risks at the expense of its allies' security.

It is more likely that Reagan and his advisers simply did not fully comprehend the ramifications of their actions for U.S.–West European relations. Nevertheless, the Europeans seemed to view those developments as the manifestation of a deliberate and not very gratifying change in U.S. policy, and they lamented the lack of a distinctly European voice on security matters. They reacted by intensifying their hedging strategy through a strengthened WEU. In an address before the Royale Institute of International Relations in Brussels, former British foreign secretary Sir Geoffrey Howe traced the WEU's revitalization to "a growing perception that a European forum was needed in which we Europeans could consult one another about our fundamental security needs." He contended that developments such as the double-zero option "underline the need for the European countries to consult more closely among themselves about their defense interests as well as with the Americans."[5]

A resurrected WEU also served as the vehicle for closer French-German cooperation on a range of military issues.[6] Paris had long urged its European colleagues to reduce their dependence on the United States for defense, and French officials now pushed their agenda in more substantial ways. Bilateral efforts with West Germany (and later a unified Germany) led to the formation of a Franco-German brigade and later to the Eurocorps.[7]

Nevertheless, as long as the looming Soviet threat existed, Western Europe's reliance on the United States for security was a fact of geostrategic life, unless the Europeans were willing to make a massive increase in their own defense efforts—and defense expenditures. The renaissance of the WEU was a hint of what might be, but no more than a hint. West European governments and publics continued to view the U.S. leadership role with profound ambivalence. On the one hand, they viewed the security tie to the United States through NATO as crucial protection against Soviet blackmail or aggression. The massive U.S. contribution to Western Europe's defense (which reached more than $130 billion a year by the late 1980s) was also both politically and economically appealing.

On the other hand, the advent of nuclear parity between the United States and the Soviet Union—a development that virtually guaranteed the destruction of the American homeland in the event of a war in Europe—made Washington's commitments under article 5 of the North Atlantic Treaty less and less credible to uneasy Europeans. And Western Europe's dependence on the United States for protection—especially when combined with Washington's insensitive, sometimes even arrogant, attitude—produced more than a small sense of humiliation. Given those conflicting pressures, the Europeans behaved inconsistently. They proclaimed their undying allegiance to NATO and their desire for continuing U.S. leadership even as their actions in strengthening the WEU and asserting at least a measure of security independence moved them in the opposite direction.

With the collapse of the Warsaw Pact and the eclipse of the Soviet conventional military threat, the pace of West European assertiveness began to increase, much to the chagrin of U.S. policymakers. The jurisdiction of the WEU received a substantial and formal boost with the adoption of the Treaty on European Union (generally referred to the Maastricht Treaty) in December 1991. Among other provisions, the Maastricht Treaty committed the European Community to an "ever closer" political union that would include common foreign and defense policies. It also called for strengthening the WEU to elaborate and implement defense decisions, thereby effectively making the WEU the defense arm of the new European Union.

Although subsequent events, including Denmark's initial rejection of the Maastricht Treaty and the virtual collapse of the concept of a unified currency, demonstrated that a significant gap remained between official proclamations of European unity and the attainment of that unity, the WEU has become increasingly visible and active. In the summer of 1992, for example, it authorized the deployment of naval units in the Adriatic to help enforce the United Nations' economic embargo against Yugoslavia. Interestingly, NATO sent ships to participate in that blockade only *after* the WEU had already acted—an unprecedented development. Moreover, the organization has taken significant steps on such mundane but important matters as implementing joint training programs and providing new impetus for cooperation in armaments procurement. Sir Dudley Smith, president of the WEU Assembly, observed in October 1993, "In the period since the Maastricht Treaty was signed there has been a striking contrast between the activity of the WEU, which has implemented most of the decisions contained in the Declaration annexed to the Treaty on European Union, and the perceived paralysis of the process of integrating Europe."[8]

### Washington's Hostility to European Initiatives

Instead of encouraging such initiatives, Washington viewed with thinly disguised hostility the various efforts to strengthen the WEU as well as other manifestations of greater European self-reliance. Bush administration officials regarded such developments as a threat to America's preeminence in transatlantic security affairs. As early as February 1991 National Security Adviser Brent Scowcroft complained that the West European governments were meeting privately on security issues and then presenting a common front to the United States.[9] A few months later the administration expended considerable diplomatic capital to pressure its European allies into agreeing to create a rapid reaction force under the auspices of NATO rather than, as France had proposed, under the European Community.

Indeed, the Bush administration sought to undermine even more limited expressions of European military initiative, such as the Franco-German Eurocorps, viewing it as the nucleus of an independent European security organization that might ultimately compete with NATO. The Eurocorps was the brainchild of President François Mitterrand of France and Chancellor Helmut Kohl of Germany. Drawing on provisions adopted at the European Community summit in late 1991 to construct a European defense "identity," the two leaders announced the formation of the Eurocorps on May 22, 1992. It would initially consist of 35,000 to 40,000 troops and be ready for deployment by 1995. The unit was designed to undertake a variety of missions, including some peacekeeping and humanitarian functions. From the standpoint of NATO partisans in Washington, however, the most worrisome mission was the mandate for action within the joint defense framework of the allies, not only in accordance with article 5 of the North Atlantic Treaty, but also in accordance with the WEU treaty. The Bush administration's uneasiness was not alleviated by the comments of German defense minister Volker Rühe that the underlying premise of the Eurocorps was to establish a European military force capable of carrying out missions in situations in which NATO would be "neither willing or able to intervene."[10]

The relationship of the Eurocorps to the WEU was close but not congruent. The Eurocorps was to be open to all members of the WEU (Belgium, in fact, decided to contribute military forces to the enterprise in early 1994, and both Spain and Luxembourg have announced their intention to do so), but it would not draw on the WEU as an institution for logistical operations. Planning and staff functions, in fact, were to be located separately in France. Moreover, as other nations were admitted to the European Union, they would be encouraged to join the Eurocorps, whether or not they decided to become members of the WEU. Such a curious and cumbersome relationship is indicative of the problems being encountered by the West Europeans as they grope toward a security identity distinct from NATO. Overlapping functions and institutions are probably an inevitable characteristic of that process.

The Bush administration's response was surprisingly intense, signifying the great hostility of pro-NATO forces in the United States to any manifestation of West European independence in security affairs. Jonathan Clarke notes that the administration reacted to the plan "with icy scorn." An administration official complained that the Eurocorps "undercut the whole American raison d'être in Europe."[11] Another unnamed high-ranking policymaker excoriated the French for having the temerity to even propose such a scheme. "The French have a second-rate nuclear deterrent, a third-rate real time intelligence capability, and a third-rate conventional army."[12] Brent Scowcroft reportedly dispatched a "strongly worded" letter to German officials demanding that they use their influence with the French to abort the agreement.

Even before the transatlantic diplomatic spat over the Eurocorps, Washington's apprehension about West European intentions was epitomized by President Bush's petulant comments at the November 1991 NATO summit meeting in Rome. Departing from the text of his prepared remarks, Bush admonished the European members of the alliance, "If, my friends, your ultimate aim is to provide independently for your own defense, the time to tell us is today."[13]

## The Clinton Administration's Policy of Limited Tolerance

To its credit, the Clinton administration has adopted a more constructive, forward-looking approach. In his speech to the NATO foreign ministers conference in December 1993, Secretary of State Christopher conceded, "Previous American administrations were ambivalent about the development of a distinct European security capability." That had all changed. "Today, the United States fully supports efforts to create a strong and effective European Security and Defense Identity (ESDI). Such an identity is a natural element of European integration."[14]

Although Christopher's remarks suggested a refreshing new attitude toward the WEU, the Clinton administration's receptivity to greater West European initiative in security matters still has limits. Christopher made it clear that the U.S. support was predicated on the belief that it would make the European Union "a more capable partner" in the pursuit of "our mutual interests." In addition to stressing the primacy of presumed mutual transatlantic interests, he qualified his endorsement of a stronger WEU in other ways. He stated, for example, that although the WEU might sometimes take "autonomous actions," when appropriate, the North Atlantic Council—NATO's governing body—would always consult on issues that affected the security of the allies. "And NATO should have full opportunity in those consultations to consider the appropriate response." Although Christopher stated that requiring those deliberations "would not contemplate an Alliance veto over WEU actions," that would be the practical effect of such a supervisory role.

Christopher expressed the wish that NATO and the WEU "work as partners." But he left little doubt about which organization was to be the senior partner. Indeed, he seemed to view a more vigorous WEU as a kind of wholly owned NATO military subsidiary that would undertake relatively minor security missions. His comments were merely the latest evidence that the Clinton administration endorses a stronger WEU only as the "European pillar" of a U.S.-dominated NATO.

That was essentially the formulation adopted at the NATO summit meeting in January 1994. The final declaration had a somewhat erratic quality as it tried to reconcile the objectives of transatlantic solidarity and a distinct West European security role. Predictably, the NATO leaders declared a "strong commitment to the transatlantic link, which is the bedrock of NATO. The continued substantial presence of United States forces in Europe is a fundamentally important aspect of that link."[15] But the language concerning the new ESDI suggested a greater assertiveness and independence on the part of the Europeans, despite the mandatory rhetorical deference to NATO's supremacy. The alliance leaders pledged "full support to the development of a European Security and Defence Identity, which, as called for in the Maastricht Treaty, in the longer term perspective of a common defence policy within the European Union, might in time lead to common defence compatible with that of the Atlantic Alliance." They went on to specifically support strengthening the WEU as the defense component of the European Union.[16]

The emphasis on defense compatibility with NATO was not an isolated comment. Throughout the declaration there was a pronounced, almost strident, effort to reiterate that the ESDI would not really be independent and that there were no fissures whatsoever in the Atlantic alliance. Thus, the emergence of the ESDI would "enable the European Allies to take greater responsibility for their common security and defence," but it would also (how, the declaration did not say) reinforce the transatlantic link. Lest anyone think otherwise, the NATO leaders affirmed, "The Alliance and the European Union share common strategic interests."[17]

Another passage that reflected the delicate balancing act of recognizing a more assertive and independent policy by the West Europeans while reiterating NATO's preeminence was one dealing with the use of alliance assets for WEU operations. After affirming NATO's willingness to make collective military assets available to the WEU, the declaration stressed support for "separable but not separate" capabilities that "could respond to European requirements" yet "contribute to

Alliance security."[18] The curious "separable but not separate" formulation occurred again in a later passage on the mechanisms of NATO-WEU cooperation.

The repeated emphasis on the allegedly congruent interests of North America and Western Europe—which reached its zenith in a flamboyant closing passage proclaiming that the two regions were "permanently committed to their common and indivisible security"—could lead a cynic to contend that the alliance leaders do protest too much. Despite the ample rhetoric about NATO's continuing primacy and the insistence that the ESDI, implemented through a stronger WEU, will in no way undermine transatlantic solidarity, the reality is almost certain to be otherwise. The development of a distinct, vigorous European security organization will inevitably erode NATO's preeminence. Like it or not, a strong WEU, by its very existence, becomes an alternative—and ultimately a competitor—to NATO. That competition may not be acrimonious, and it may not lead to military estrangement between the United States and Western Europe in the short term, but the two "pillars" of the Western alliance will begin to separate. The task for American policymakers is to accept Western Europe's impending security independence with maximum grace and, indeed, to help accelerate the process.

**Washington Should Encourage the WEU**

More flexible and progressive U.S. thinking about the WEU is needed. Even at the time NATO was created, some U.S. policymakers favored a more limited American security relationship with Western Europe than the North Atlantic Treaty implied. Such prominent State Department officials as George Kennan and Charles E. Bohlen believed that there was no need for a comprehensive transatlantic military alliance and advocated instead an agreement between the United States and the Brussels Pact.[19] That type of limited partnership, sufficient to protect America's bona fide security interests in Europe without entangling the United States in purely regional disputes and conflicts, would be even more appropriate today than at the dawn of the Cold War.

Nine of the 12 (soon to be 16) nations in the European Union, including all of Western Europe's most significant military powers, are already members of the WEU. Collectively, the WEU states have nearly 2 million active duty military personnel, and two of the signatories,

Britain and France, have modest nuclear arsenals. In short, the WEU has significant and credible military capabilities. It is also the logical security arm of the European Union as that association strives to develop a common security policy.

NATO partisans invariably respond that the WEU could never be an adequate substitute. Josef Joffe contends that the WEU would face even greater credibility problems than NATO would in dealing with European security matters. He asks, "If NATO has lost its rationale along with its rival, why should the EC/WEU do better?"[20] Jeffrey Simon dismisses suggestions that the WEU replace NATO on the grounds that not only does the WEU omit Greece and "non-EC flank members" but, more damning, it "omits the trans-Atlantic link with the United States and Canada."[21] Still other NATO advocates argue that a powerful WEU would again split Europe into antagonistic security blocs, because the East European states would be excluded.

Although such criticisms must be taken seriously, they are hardly compelling reasons for not making the WEU a successor to NATO. Joffe, for example, ignores a crucial difference between the two organizations. NATO was created as one superpower-led bloc to blunt the expansionist drive of the rival superpower. The demise of the Soviet threat does indeed erode the rationale for such a far-reaching alliance. But the collapse of the Soviet Union does not mean that Russia will forever be absent from the ranks of Europe's great powers. Russia no longer has a credible claim to the title of superpower, but even in its current weakened state it is still a factor in Europe's strategic affairs, and with its anticipated political and economic recovery, it may again become the single strongest power on the Continent—or at least compete with Germany for that status. Under such conditions, the WEU would be the logical counterweight to Russia's military strength in a reasonably stable European balance of power. It would have enough military potential to deter widespread Russian expansionist ambitions without being (in contrast to a U.S.-led NATO) so powerful that it would be viewed in Moscow as a menace to Russia's vital interests. What Joffe fails to realize is that the rationales for NATO and the WEU are fundamentally different in a post–Cold War setting. Among other things, a security role for the latter is more plausible in the long term because of a similarity of interests among the West European states that is increasingly absent in the artificial concept of a transatlantic community.

Two caveats need to be made, however, about a strong role for the WEU. First, although some expansion of WEU membership is probably both inevitable and desirable, expansion into Central (much less Eastern) Europe could entail some of the same risks as a similar enlargement of NATO.[22] Moscow might not be quite as uneasy about WEU intrusion into that region as it would be about a NATO presence, but it would hardly be pleased.

Second, and more important, an enlarged WEU must be truly independent, not merely a surrogate for NATO. Thus, the proposal by defense analyst Steven L. Canby to incorporate the Visegrad countries into "NATO's newly recognized junior partner, the military arm of Maastricht, the West European Union," rather than NATO itself, while superficially appealing, contains insidious dangers. Although Canby contends that such a maneuver "stings Russia less," that is doubtful if Moscow has reason to view the WEU as nothing more than NATO's puppet. Indeed, Russia has already criticized the WEU's offer of associate membership to nine Central and East European countries. A unnamed high-level U.S. official admitted that the main reason for Moscow's opposition was that "Russia saw the measure as a way of ushering countries like Poland and the Czech Republic into NATO through the back door."[23] Russian officials had reason for such suspicion. For example, Latvia's justice minister, Egils Levits, told reporters that his government regarded associate membership with the WEU as "entrance to NATO via detour."[24]

The second danger of using the WEU as a surrogate for NATO is identified by Canby himself, although he regards it as a benefit rather than a problem.

> Militarily, the WEU approach extends the American nuclear deterrent less visibly but in a way substantially no different from what was developing in NATO before the Soviet collapse.... The use of Russian nuclear weapons against the Visegrad countries would certainly bring to the fore the American deterrent. Should the U.S. demur at the moment of truth, there are always French nuclear weapons to trigger events.[25]

Canby is absolutely right—if the WEU is merely NATO's European pillar. From the standpoint of American interests, the whole point of getting the WEU to supplant NATO is to reduce the likelihood of entangling the United States in an East European conflict. But if the two organizations are linked like Siamese twins, the danger of entanglement will be as great as, or conceivably even greater than, under the current system. Indeed, as Canby acknowledges, France (or Western Europe's other nuclear power, Britain) could force the United States into a nuclear confrontation with Russia against Washington's will. The only mystery is why Canby or any other analyst would regard such a prospect favorably.

The underlying problem with Jeffrey Simon's analysis as well as other hostile critiques of the WEU is that NATO's defenders essentially demand that any successor organization embody all of NATO's characteristics, including the transatlantic link to the United States and the same geographic coverage. But is such a comprehensive alliance needed in post–Cold War Europe? It may instead be sufficient to have several organizations with different mandates and missions. The WEU should be capable of protecting Western Europe from external attack, and the members would have an obvious self-interest in defending the security of their region. The CSCE could serve not only as a continental forum for discussing broader problems but also more specifically as a mechanism for resolving some quarrels in Eastern Europe before they explode into armed conflict.

Perhaps a limited security association among the Central and East European nations—roughly an Eastern analog to the WEU—would also emerge. That would provide some insurance against the possibility of a revanchist Russia without being as threatening to Moscow as the expansion of a U.S.-led NATO—or even the WEU—to the borders of the Russian Federation. There were serious discussions during 1992 and 1993 among various East European governments, especially those of the Visegrad powers, about establishing a mutual defense pact for their region. Ironically, one of the earliest effects of the Partnership for Peace, and the lure of eventual NATO membership, was to retard, and perhaps abort, that process. The Czechs, in particular, now are apparently pinning nearly all their security hopes to their country's early admission to NATO.[26] Instead of pushing the Partnership for Peace, Washington should encourage the East European states to look after their own defense needs, including the creation of a regional alliance, if they deem that step appropriate.

Without the menace posed by a powerful state bent on continental domination, there is little need for a "one size fits all" European (much less transatlantic) military alliance. Above all, the various issues, problems, and tradeoffs are properly matters for the Europeans to work out for themselves. The precise characteristics of post-NATO security organizations would depend on numerous factors, including the goals of the key European powers, the burdens and risks those powers were willing to assume, and the nature of the threat environment that emerges in the coming years. In any case, there is no compelling reason for American officials to insist on a comprehensive post–Cold War European security architecture designed by Washington.

## Misguided Efforts to Keep Europe Dependent

NATO's defenders in the United States seem fearful that the Europeans will take responsibility for their own defense. Ronald D. Asmus, Richard L. Kugler, and F. Stephen Larrabee, for example, fret that as new countries "enter the EC, they will have the option of joining the Western European Union. Thus, they will have security guarantees from Europe's key powers and Washington's major NATO allies. A situation in which a country like Germany would extend a security commitment to Poland through the WEU, but not through NATO, could destroy the Atlantic alliance."[27] Similarly, Sen. Richard Lugar (R-Ind.) warns that "if NATO does not deal with the security problems of its members, they will ultimately seek to deal with these problems either in new alliances or on their own."[28] According to the logic of Atlanticists, that would be an unfortunate development, despite the significant financial savings to American taxpayers and the reduction of America's exposure to the risk of entanglement in local conflicts.

The most frequently cited reason why greater European self-reliance would be undesirable is that it would lead to the "renationalization" of Western Europe's defenses and the reignition of the national rivalries that have rent the Continent in the past. The possibility of renationalization has some merit, and the potential consequences must be taken seriously. Indeed, in its most extreme form the process could lead to the acquisition of nuclear weapons by Germany and other states, with dangerously destabilizing implications. There are reasons to believe, however, that the future of Western Europe will not follow that nightmare scenario.

Lugar and others who invoke the renationalization specter ignore the extensive institutional patterns of cooperation that have developed in Western Europe over a period of nearly half a century. Although the

West Europeans may not achieve the degree of economic and political unity contemplated in the Maastricht Treaty, the economies of the member states of the European Union are already tightly linked. Throughout the political and security realms, there are also far more areas of common interest than there are areas of potential conflict. The disagreements that do exist are often more about implementation than about crucial matters of substance. Those extensive bonds will not rupture merely because the United States withdraws its military forces from Europe and devolves security responsibilities to its European allies.

One might also wonder when the countries of Western Europe will be sufficiently stable and democratic to be trusted to manage their own security affairs, if they are not now, after nearly a half century of peace and cooperation. The notion that U.S. leadership and an American troop presence are the only reliable guarantees that the West Europeans will not revert to bloodletting is overstated—and more than a little condescending. The nations of Western Europe have significant mutual political, economic, and security interests in keeping their national rivalries within bounds. And if those mutual interests do not prove sufficient to prevent rampant renationalization, the mere existence of NATO is not likely to be an effective restraint. Renationalization, although a worrisome possibility, is by no means a foregone conclusion, and it is not a compelling justification to maintain NATO and the U.S. military presence on the Continent in perpetuity.

Those who insist that there is no substitute for U.S. leadership invariably point to the inaction of the European Union in the Yugoslavian crisis as confirmation. According to the conventional wisdom, in the absence of U.S. leadership the Europeans have displayed cowardice and ineptitude, with disastrous results. That indictment is, for the most part, unfair.

There are two reasons why the West Europeans have been reluctant to intervene militarily in Yugoslavia, and neither reflects cowardice or ineptitude. The first is a healthy appreciation for the intensity of the ethnic hatreds throughout the Balkans and the enormous potential for getting bogged down in a bloody, pointless military adventure. (A similar caution may explain why some of the West European powers are ambivalent, at best, about expanding NATO into Eastern Europe.)

The other reason for the hesitation about Yugoslavia is the European Union's inexperience in taking responsibility for managing security problems. That inexperience is largely the result of the domination the United States has exercised since the late 1940s as the leader of NATO. Even the "burden-sharing" complaints that surfaced periodically during the Cold War reflected a desire that the Europeans pay more of the costs of policies adopted by Washington, not a willingness to accord the allies a larger role in decisionmaking. Ronald Steel points out that "for all practical purposes Europe's defense and diplomacy were run from Washington."[29]

As the West Europeans have encountered difficulties in dealing with the Yugoslavian crisis, the old habit of looking to Washington for a solution has reasserted itself. Increasingly, the European Union governments want the best of both worlds: decisive action without having to bear more than a modest portion of the costs and risks. Such an attitude is one of the more insidious legacies of a dependent mentality—the avoidance of responsibility even when significant interests are at stake.

The longer Washington insists that NATO eclipse the WEU or any other competing organization, the longer it will take the European states to again become competent security actors. In addition to reducing America's risk exposure in parochial conflicts, phasing out NATO would have the salutary effect of ending Western Europe's artificial and unhealthy military dependence on the United States.

That realization has belatedly begun to dawn on some West European leaders and policy experts. Unsurprisingly, the trend appears to be strongest among the French, who even during the Cold War had a more realistic view of U.S. policy than did most of their European colleagues. The Gaullist faction assumed that U.S. and West European interests did not always coincide, and that there were limits to the risks Washington was willing to take and the sacrifices it was willing to make on behalf of its allies. A consistent theme of French policy under both Charles de Gaulle and his successors was a quest for a more distinctly European security identity that would reduce Western Europe's reliance on the United States.

Such thinking has persisted in the post–Cold War years. Viewing the results of the January 1994 NATO summit, Pierre Lellouche, a senior scholar at the French Institute for International Relations, concluded that the United States was in no mood to become involved in Bosnia-style conflicts or crises. Consequently, "Western Europe now stands alone in the face of destabilization of the eastern part of the continent." Furthermore, as long as the only security mechanism is a U.S.-dominated NATO, "Europeans are condemned to becoming impotent spectators of the convulsions which threaten them directly."[30]

Although one might take issue with Lellouche's assumption that the U.S. political elite (as opposed to the American public) is determined not to see the United States entangled in Eastern Europe's conflicts, his concern about the adverse consequences of Europe's dependence on Washington is well-founded. Instead of offering the NATO allies new assurances of America's undying commitment to every aspect of the Continent's security, Washington should stoke the embers of European realism by stressing that there are rigorous limits to the risks we are willing to take and the costs we are willing to bear. The time has come—indeed it is long overdue—to acknowledge that while European and American security interests may overlap, they are by no means identical. The policy implication of such candor is that the West Europeans would be wise to create robust security organizations of their own.

What Lugar and other NATO partisans in the United States actually fear is not that the West Europeans will be unable to manage their own affairs, but that they will be too successful and therefore have no further need of the United States. It is the nightmare of possible U.S. exclusion from Europe that animates American Atlanticists. That fear was clearly evident in former UN ambassador Jeane Kirkpatrick's lament about efforts to promote greater political and economic unity among the West Europeans: "The most important conversations about the future organization of the Western world are taking place without any U.S. participation."[31]

It is not coincidental that Kirkpatrick and others who express fears about the West European states' seeking to exclude the United States in important matters are also among the loudest proponents of new missions for NATO. Maintaining NATO is crucial, they argue, because it is the only significant European institution in which the United States has a seat at the table. (Indeed, Washington occupies the seat at the head of the table.) In contrast, the United States has no way of participating in decisions reached by the European Union. NATO leadership, and the security ties with Western Europe it symbolizes, is the principal "leverage"—perhaps the only remaining meaningful leverage—Washington has to guarantee that its political and economic interests on the Continent will be respected by the major West European powers.

But Washington's ability to translate its dominance in the transatlantic security relationship into comparable influence on such nonmilitary matters as trade, monetary policy, and out-of-area diplomatic initiatives was overrated even during the Cold War.[32] The allies typically made just enough (frequently grudging) concessions on those issues to prevent an open breach in the alliance.

Indeed, the European countries were not above using "reverse leverage" on the United States when it suited their economic interests. A report by the British Broadcasting Corporation about an incident between the United States and Britain in 1985 indicates that America's closest ally was apparently willing to risk rupturing Western security solidarity over a surprisingly limited issue. According to the BBC, the government of Margaret Thatcher threatened to sabotage U.S.-Soviet strategic arms talks if the United States proceeded with plans to prosecute British Airways. The state-run airline, earmarked for privatization by the Thatcher government, was accused of engaging in illegal practices to force Laker Airways, a smaller rival, out of business. A successful prosecution could have seriously damaged the value of British Airways, reducing its appeal to private investors, perhaps even jeopardizing the privatization plan. The BBC reported that Thatcher openly threatened to withdraw London's support for the U.S. position on ballistic missile defense and other issues at the upcoming Strategic Arms Limitation Talks in Moscow unless the Reagan administration called off the investigation and impending prosecution.[33] That Washington's closest NATO ally would play such diplomatic hardball—especially on a relatively minor issue—suggests that the U.S. leverage thesis was overblown even in a Cold War setting.

If the United States had difficulty exercising effective leverage when the importance of its military protection to the West Europeans was quite substantial, it is certain to encounter even greater problems in the post–Cold War era. The value of the U.S. security guarantee has depreciated markedly with the demise of the Soviet threat. Although the West European nations might wish to retain defense ties with the United States as a long-term insurance policy and a valuable financial subsidy, U.S. protection is no longer widely viewed as irreplaceable. If NATO enthusiasts in the United States believe that the alliance commitment can be parlayed into reliable U.S. influence on economic and political matters, they may be in for a rude awakening.

Secretary of Defense William Perry tacitly conceded the meager extent of U.S. leverage when he pleaded with Congress to refrain from making demands for greater burden sharing. (The House of Representatives had passed an amendment to the defense authorization bill for fiscal year 1995 threatening to reduce the number of U.S. troops stationed on the Continent if the European members of NATO did not pay a greater portion of their direct expenses.) "It is unrealistic to expect that our allies can be convinced to increase host nation support and other contributions," Perry stated.[34]

Rep. Barney Frank (D-Mass.), the sponsor of the amendment, dismissed Perry's objections as "nonsensical." Frank contended that "if it [the troop presence] is really important, the Europeans should be willing to pay."[35] But that is precisely the point; the U.S. deployment of forces, while desirable from the perspective of the West Europeans, is not so crucial that they are willing to make significant financial sacrifices. If U.S. officials fear that the allies are unwilling to placate the United States on this narrow burden-sharing issue—which involves at most a few billion dollars a year—expectations that those countries will capitulate on trade and monetary matters involving far greater stakes are more unwarranted.

### Ending the Atlanticist Fantasy

Instead of frantically trying to preserve NATO, Washington should encourage the creation of "Europeans-only" security organizations as successors. The North Atlantic alliance was established in response to conditions unique to the Cold War. It was plausible to argue that only an alliance led by a superpower could neutralize the military threat posed by another superpower. But that consideration no longer applies. The members of the European Union have more than sufficient economic, political, and military resources to deal with lesser threats to their security—including the nebulous menace of instability among their eastern neighbors. The EU nations, with the four new members slated for admission, will collectively have more than 375 million people; a collective economy of $7.5 trillion a year; and well-trained, well-equipped military forces. Those nations also have a more substantial stake than does the United States in developments in Eastern Europe.

Expanding or even preserving NATO in a post–Cold War setting is based on the fiction that U.S. and European security interests are not only similar but virtually congruent. The notion of a transatlantic security community was exaggerated even during the Cold War. Although the United States and its alliance partners faced a common adversary, there were still ample occasions for serious policy disagreements. Signs of diverging interests grew increasingly evident as the Cold War dragged on. That was especially true with regard to "out of area" issues, such as Middle East policy and attitudes toward left-wing regimes or insurgencies in Central America, but there was also intra-alliance friction over relations with the Soviet Union itself. Disputes about trade relations with the Eastern bloc, technology transfers, the role of nuclear weapons in NATO's defense strategy, and other matters became surprisingly acrimonious in the 1970s and 1980s.[36]

The concept of a transatlantic security community in the post–Cold War era is even less realistic. Owen Harries, editor of the *National Interest*, asks the pertinent question of whether "the West" exists as a meaningful concept any longer.

> Proposals for what amount to a new NATO are based on a most questionable premise: that "the West" continues to exist as a political and military entity. Over the last half century or so, most of us have come to think of "the West" as a given, a natural presence and one that is here to stay. It is a way of thinking that is not only wrong in itself, but is virtually certain to lead to mistaken policies. The political "West" is not a natural construct but a highly artificial one. It took the presence of a life-threatening, overtly hostile "East" to bring it into existence and to maintain its unity. It is extremely doubtful whether it can now survive the disappearance of that enemy.[37]

Harries points out that despite the many common roots of their civilizations, there was little concept of political unity between the United States and Europe before World War II. Indeed, each side tended to view the other with either suspicion or contempt. Despite the rhetoric of solidarity, the actions of the United States and the major powers of Western Europe since the end of the Cold War have again suggested more disharmony and rivalry than unity.

> As soon as the Soviet Union disintegrated, what we immediately started to hear propounded on both sides of the Atlantic—and what we were still hearing from President Clinton during the July [1993] Tokyo summit—was a tripartite or

tripolar version of the world, with Europe and the United States again constituting not one but two separate sides of the triangle, and with Japan/Asia as the Third. Far from stressing the continuing existence of "the West," once free of a Soviet threat many Europeans immediately began anticipating, often with ill-concealed glee, a post-Maastricht United Europe that would supplant the United States as the dominant economic—and ultimately political—force in the world.[38]

Developments in a number of policy areas have generally confirmed Harries's analysis. Certainly the bitter confrontations between the United States and various West European countries during the Uruguay Round of the General Agreement on Tariffs and Trade (GATT) were hardly symptomatic of Western solidarity. Equally caustic exchanges on economic and fiscal policy—including intense U.S. efforts to pressure the German government to lower its interest rates—likewise indicated friction rather than harmony. And the nasty scraps over such security issues as the formation of the Eurocorps have already been noted. Warren Christopher's jarring comments in October 1993 that Washington for too long had had a "Eurocentric attitude," and that "Western Europe is no longer the dominant area of the world," were symbolic of the fraying fabric of meaningful transatlantic unity.[39]

Despite the grandiose rhetoric of some Atlanticists, the United States is not a European power. America is an external power that has some economic and security interests in Europe. That is a subtle but very important difference. In a post–Cold War setting it is likely that the interests of the United States and those of Western Europe will coincide on some issues, be markedly different on others, and perhaps even be in direct conflict on still others.

The appropriate way of handling a situation that is far more complex and fluid than it was during the era of Cold War bipolarity is to establish a more limited and flexible security relationship with the nations of Western Europe. Such a relationship need not—in fact, should not—be enshrined in elaborate treaties that attempt to codify cooperation on a wide range of diverse issues. A worthwhile step would be to replace NATO with a security coordination council. In a relatively low-threat environment, the council would serve primarily as a forum for discussing political and military issues of mutual concern. That forum could quickly become a mechanism for consultations if a significant security problem began to emerge. It could even become a strategy planning group should a crisis evolve that represented a significant threat to the security of countries on both sides of the Atlantic. Such an institutional arrangement would not require a lengthy and solemn treaty of alliance, much less an extensive bureaucratic, political, and military infrastructure.

When formal agreements are needed at all, they ought to be kept as direct and Spartan as possible, usually focusing on a single topic of mutual concern, or at most a small number of related issues. Indeed, whenever possible the United States should opt for informal understandings rather than binding, formal accords. The guiding principle should be an insistence on maximizing America's policy options and preserving its decisionmaking autonomy.

That approach would allow cooperation when common interests were threatened but not entail awkward or undesirable obligations when such common interests and objectives were absent. It would also be more flexible and responsive to a rapidly changing international system than a formal alliance would be. Commitments that might make sense under one set of conditions can become irrelevant, counterproductive, or even dangerous when conditions change. Formal, permanent alliances such as NATO fail to take that reality of international politics into account.

U.S. policymakers should firmly reject proposals to transform NATO into an organization for performing missions for which it was not designed. Instead, they must come to recognize that the alliance has outlived its usefulness and that entirely new security arrangements are needed in post–Cold War Europe. Those arrangements ought to be directed by Europeans and tailored to European concerns and requirements. The campaign to enlarge NATO is a desperate and ill-advised attempt to resuscitate a moribund Cold War era institution.

In addition to establishing a more limited and focused security relationship with Western Europe, the United States should seek ways to expand its economic ties with the entire Continent. The latter objective needs to include not only a more effective relationship with the European Union—as well as upgraded bilateral links to the key member states of that association—but also more coherent policies toward the nations of Eastern Europe and the former Soviet Union. Since the end of the Cold War, an assortment of tensions has arisen in

America's economic relations with European countries, both West and East. Many of those problems would be avoidable if Washington were to adopt more rational policies.

Other problems are more difficult to resolve and require intensive bilateral or multilateral efforts. The need for U.S. officials to pay greater attention to burgeoning U.S.-European quarrels on the economic front is becoming more urgent. Such disputes are far more likely in the short and medium term to have an adverse impact on American interests than is the remote danger that a serious military threat might emerge.

Unfortunately, because of the reliance on NATO, Washington's institutional relationships with Europe are predominantly military in nature. That arrangement is ill-suited for dealing with economic problems, despite the assumption of America's NATO enthusiasts that transatlantic security ties and the presence of American forces in Europe will translate into effective U.S. economic leverage. Jonathan Clarke makes the more practical suggestion that Washington "move its institutional arrangements with Europe away from security issues to economic matters on which it is absolutely critical that the United States try to create relationships of institutional intimacy."[40] Clarke cites the extraordinary difficulties encountered in negotiating the Uruguay Round of the GATT, especially the bitter disputes between U.S. and EU representatives on such matters as agricultural subsidies, as an omen of a conflict-ridden future without such institutional relationships. He also warns that if the European Union and the signatories to the North American Free Trade Agreement "were to develop into antagonistic trading blocs, the consequences for global prosperity would be catastrophic."[41]

Instead of trying to preserve NATO, it would be more useful for U.S. leaders to propose the creation of a transatlantic economic council that would focus on preventing or resolving U.S.–West European economic disputes. The mandate for the council should be to promote and protect the most open commercial system possible. Such an institutional structure would be especially helpful in addressing problems in their early stages rather than allowing them to fester and become more disruptive. The present approach is little more than a patchwork of ad hoc consultations—usually on a bilateral basis and often taking place only when a disagreement has reached the point where recriminations are flying back and forth across the Atlantic. The ugly exchange of accusations that marked the negotiations of the Uruguay Round is precisely the kind of situation that a transatlantic economic council would strive to prevent.

Establishing an economic council and creating a new security relationship with the WEU require new thinking on the part of an intellectually sclerotic American foreign policy establishment. It would mean overcoming the twin fallacies that NATO is an alliance for all seasons and that NATO is the sole effective mechanism for promoting and protecting U.S. political and economic interests in Europe. The post–Cold War era cries out for a more flexible and creative U.S. policy toward Europe. It is time to stop treating NATO as an icon that must be preserved even if it means undertaking unrewarding and dangerous new missions.

# Notes

## Introduction

1. Examples include Jeffrey Simon, "Does Eastern Europe Belong in NATO?" *Orbis* 37 (Winter 1993): 21–35; Ronald D. Asmus, Richard L. Kugler, and F. Stephen Larrabee, "Building a New NATO," *Foreign Affairs* 72 (September–October 1993): 28–40; Zalmay M. Khalilzad, "Extending the Western Alliance to East Central Europe," RAND Corporation Issue Paper, May 1993; Zbigniew Brzezinski, "A Bigger—and Safer—Europe," *New York Times*, December 1, 1993, p. A23; Zbigniew Brzezinski, "The Premature Partnership," *Foreign Affairs* 73 (March–April 1994): 67–82; James A. Baker III, "A NATO Carrot to Solidify Reform," *Washington Times*, December 6, 1993, p. A18; and William S. Cohen, "Expand NATO Step by Step," *Washington Post*, December 7, 1993, p. A25. A more cautious and nuanced argument for limited NATO expansion can be found in Charles L. Glaser, "Why NATO Is Still the Best: Future Security Arrangements for Europe," *International Security* 18 (Summer 1993): 13–14.

2. *Congressional Record*, 103rd Cong., 1st sess., January 27, 1994, pp. 267, 273–74.

3. For examples, see Rowland Evans and Robert Novak, "Ghost of Yalta," *Washington Post*, November 18, 1993, p. A23; "Visions of NATO and Visions of Russia," editorial, *Washington Times*, November 5, 1993, p. A22; and Center for Security Policy, "Yalta II: Western Moscow-Centrism Invites New Instability in Former Soviet Empire," Decision Brief no. 93-D101, December 3, 1993.

4. Steve Vogel, "U.S. Proposes NATO 'Partnerships' for Former Warsaw Pact Nations," *Washington Post*, October 21, 1993, p. A22.

5. Quoted in Douglas Jehl, "Ukrainian Agrees to Dismantle A-Arms," *New York Times*, January 13, 1994, p. A6.

6. See the comments of Field Marshal Sir Richard Vincent, chairman of NATO's Military Committee, in David White, "Military Chief Fears for NATO Expansion," *Financial Times*, November 10, 1993, p. 3. The comments of Boris Biancheri, Italy's ambassador to the United States, indicated that his government held similar views. Andrew Borowiec, "Italian Envoy Urges NATO Not to Look East," *Washington Times*, December 2, 1993, p. A11.

7. Quoted in "NATO Chief Sees Benefits in Admitting E. Europeans," *Chicago Tribune*, November 27, 1993, p. 18.

8. Quoted in John F. Harris, "U.S. Role in Bosnia to Increase," *Washington Post*, July 18, 1994, p. A1.

9. Richard G. Lugar, "NATO: Out of Area or Out of Business: A Call for U.S. Leadership to Revive and Redefine the Alliance" (remarks delivered to the Open Forum of the U.S. Department of State, August 2, 1993), p. 7.

151

## Chapter 1

1. Vaclav Havel, "New Democracies for Old Europe," *New York Times*, October 17, 1993, p. E17.

2. Barry Schweid, "Baltics OK Joint Exercises with NATO," *Washington Times*, October 28, 1993, p. A16; "Romania Trains to Join NATO," Armed Forces Information Services *Current News*, July 27, 1993, p. 16; and Gary Miller, "Romania to Seek Place in NATO When Time Is Right, General Says," *European Stars and Stripes*, November 18, 1993, p. 3.

3. Quoted in Celia Woodard, "Hungary Winces as West Defers Its NATO Membership," *Christian Science Monitor*, October 28, 1993, p. 3. See also Pavel Bratinka, "The Challenge of Liberation: The View from the Czech Republic," Svetoslav Bombik, "Returning to Civilization: The View from Slovakia," Jerzy Marek Nowakowski, "In Search of a Strategic Home: The View from Poland," and Tamas Waschler, "Where There's a Will . . . : The View from Hungary," in *NATO: The Case for Enlargement* (London: Institute for European Defence and Strategic Studies, 1993).

4. Paul Bedard, "Walesa to Clinton: Russians Coming," *Washington Times*, July 7, 1994, p. A11.

5. James A. Baker III, "A NATO Carrot to Solidify Reform," *Washington Times*, December 6, 1993, p. A18.

6. Ronald D. Asmus, Richard L. Kugler, and F. Stephen Larrabee, "Building a New NATO," *Foreign Affairs* 72 (September–October 1993): 28.

7. Jeffrey Simon, "Does Eastern Europe Belong in NATO?" *Orbis* 37 (Winter 1993): 21.

8. Ibid., p. 25.

9. Charles L. Glaser, "Why NATO Is Still the Best: Future Security Arrangements for Europe," *International Security* 18 (Summer 1993): 13.

10. Gary L. Geipel, "Expanding NATO: Good for Europe *and* Good for the U.S.," Hudson Briefing Paper no. 161, February 1994, p. 9.

11. Simon, pp. 32, 34. For a discussion of the creation of the North Atlantic Cooperation Council and the role it has played, see William Yerex, "The North Atlantic Cooperation Council: NATO's *Ostpolitik* for Post–Cold War Europe," in *NATO's Eastern Dilemma*, ed. David G. Haglund, S. Neil MacFarlane, and Joel J. Sokolsky (Boulder, Colo.: Westview, 1994), pp. 181–94.

12. Edward L. Rowny, "NATO and the Difference between Eastern and Central Europe," *Washington Times*, March 15, 1994.

13. Richard G. Lugar, "NATO: Out of Area or Out of Business: A Call for U.S. Leadership to Revive and Redefine the Alliance" (remarks delivered to the Open Forum of the U.S. Department of State, August 2, 1993), p. 10.

14. Barry Schweid, "Aid Boost to Kazakhstan Is Linked to Weapons, Oil," *Washington Times*, October 25, 1993, p. A11.

15. That position is endorsed by the *New York Times* as well. "Include Russia, Too," editorial, January 9, 1994, p. E20.

16. U.S. Senate Committee on Foreign Relations, *The Vandenberg Resolution and the North Atlantic Treaty: Hearings Held in Executive Session on S. Res. 239* (80th Cong., 2d sess., 1948) *and Executive L* (81st Cong., 1st sess., 1949), Historical Series, 1973, p. 104.

17. Simon, p. 33. Edward Rowny makes a similar proposal, outlining an interim

membership "that grants all rights except those under Article V." Rowny, "NATO and the Difference."

18. Asmus, Kugler, and Larrabee, pp. 35–36.

19. Quoted in David Ottaway and Peter Maass, "Hungary, NATO Grope toward New Relationship," *Washington Post*, November 17, 1993, p. A32.

20. Quoted in Ann Devroy and Daniel Williams, "Clinton Boosts A-Arms Pact in Ukraine," *Washington Post*, January 13, 1994, p. A1.

21. "Memorandum of the Government of the Slovak Republic on Joining Partnership for Peace," press release, Bratislava, February 1, 1994, p. 1. Officials in Latvia and Estonia have emphasized the same points. "Minister on 'Partnership for Peace' Expectations," February 4, 1994, "Foreign Affairs Minister on Plan," February 4, 1994, and "Official: Bilateral Ties with NATO States Needed," February 7, 1994, *Foreign Broadcast Information Services Daily Report: Central Eurasia*, SOV 94–025, February 7, 1994, pp. 70, 72; and Lennart Meri, "Estonia, NATO and Peacekeeping," *NATO Review*, April 1994, pp. 6–8.

22. Interview on National Empowerment Television, January 20, 1994.

23. For discussions of this point, see Lawrence S. Kaplan, *NATO and the United States: The Enduring Alliance* (Boston: Twayne, 1988), passim; Christopher Layne, "Atlanticism without NATO," *Foreign Policy* 67 (Summer 1987): 22–45; Ted Galen Carpenter, "United States NATO Policy at the Crossroads: The Great Debate of 1950–1951," *International History Review* 8 (August 1986): 389–415; and Ted Galen Carpenter, "Competing Agendas: America, Europe, and a Troubled NATO Partnership," in *NATO at 40: Confronting a Changing World*, ed. Ted Galen Carpenter (Lexington, Mass.: Lexington Books, 1990): 29–42.

24. Christopher Layne, "Continental Divide; Time to Disengage in Europe," *National Interest* 13 (Fall 1988): 13–27; and Layne, "Atlanticism without NATO."

25. Examples of such reasoning can be found in W. W. Rostow, *The Diffusion of Power, 1957–1972* (New York: Macmillan, 1972), pp. 396, 594; and Henry A. Kissinger, *White House Years* (Boston: Little, Brown, 1979), pp. 938–49.

26. U.S. Senate Committee on Foreign Relations, *North Atlantic Treaty: Hearings on Executive L*, 81st Cong., 1st sess., 1949, pp. 47, 213, 217, 321.

27. John M. Goshko, "NATO Pledges Increased Cooperation with Countries of Eastern Europe," *Washington Post*, June 7, 1991, p. A17.

28. Patrick E. Tyler, "Pentagon Imagines New Enemies to Fight in Post–Cold War Era," *New York Times*, February 17, 1992, p. A1; and Patrick E. Tyler, "7 Hypothetical Conflicts Foreseen by Pentagon," *New York Times*, February 17, 1992, p. A8.

29. Jim Hoagland, "When Russia and Germany Talk . . . ," *Washington Post*, May 17, 1994, p. A17. For a rare critical view of the wisdom of extending the alliance eastward by a German scholar, see Holger H. Mey, "New Members—New Mission: The *Real* Issues behind the New NATO Debate," *Comparative Strategy* 13 (1994): 223–29.

30. Bruce Clark, "Rühe Raises Polish Hopes over NATO," *Financial Times*, July 19, 1994, p. 2.

31. Quoted in Brooks Tigner, "NATO, Russia to Probe Bilateral Ties," *Defense News*, February 28–March 6, 1994, p. 3.

32. Quoted in ibid.

33. Edward Mortimer, "Better Part of Valour," *Financial Times*, February 2, 1994, p. 10.

34. Warren Christopher, "Strengthening the Atlantic Alliance through a Partnership for Peace" (remarks at the North Atlantic Council Ministerial Meeting, NATO Headquarters, Brussels, December 2, 1993), U.S. Department of State *Dispatch*, December 13, 1993, pp. 857–59.

35. Ibid.

36. Geipel, p. 2.

37. Partnership for Peace: Invitation and Framework Document, issued by the Heads of State and Government, meeting of the North Atlantic Council, Brussels, January 10–11, 1994, U.S. Department of State *Dispatch* 5, supplement no. 1 (January 1994): 5.

38. That point became even more evident after Clinton's speeches in Latvia and Poland in July 1994. White House, Office of the Press Secretary, "Remarks by President Clinton and President Ulmanis at Freedom Monument," Riga, Latvia, July 6, 1994, p. 4; and White House, Office of the Press Secretary, "Remarks of the President to the Sejm," Warsaw, Poland, July 7, 1994, p. 3.

39. Thomas W. Lippman, "NATO Peace Partnership's New Look: A Protective Shield against Moscow," *Washington Post*, February 8, 1994, p. A11.

## Chapter 2

1. U.S. Senate Committee on Foreign Relations, *North Atlantic Treaty: Hearings on Executive L*, 81st Cong., 1st sess., 1949, p. 10.

2. *Public Papers of the Presidents of the United States: Harry S Truman, 1949* (Washington: Office of the Federal Register, 1964), pp. 196–97.

3. In his account of the creation of NATO, Truman repeatedly emphasized the dire Soviet threat. Harry S Truman, *Years of Trial and Hope* (Garden City, N.J.: 1956), pp. 241–51.

4. Quoted in Henry Kissinger, *White House Years* (Boston: Little, Brown, 1979), p. 80.

5. Benjamin C. Schwarz, "NATO at the Crossroads: Reexamining America's Role in Europe," RAND Issue Paper, January 1994, p. 1.

6. Christopher Layne and Benjamin Schwarz, "American Hegemony—Without an Enemy," *Foreign Policy* 92 (Fall 1993): 5.

7. U.S. Senate Committee on Foreign Relations, *The Vandenberg Resolution and the North Atlantic Treaty: Hearings Held in Executive Session on S. Res. 239* (80th Cong., 2d sess., 1948) *and Executive L* (81st Cong., 1st sess., 1949), Historical Series, 1973, p. 221. Emphasis added.

8. For further discussions, see Ted Galen Carpenter, "United States NATO Policy at the Crossroads: The 'Great Debate' of 1950–1951," *International History Review* 8 (August 1986): 389–415; and Ted Galen Carpenter, "Competing Agendas: America, Europe, and a Troubled NATO Partnership," in *NATO at 40: Confronting a Changing World*, ed. Ted Galen Carpenter (Lexington, Mass.: Lexington Books, 1990), pp. 29–44.

9. Charles L. Glaser, "Why NATO Is Still Best: Future Security Arrangements for Europe," *International Security* 18 (Summer 1993): 8, 9.

10. Anthony Lake, "Bosnia: America's Interests and America's Role" (remarks at Johns Hopkins University, Baltimore, April 7, 1994), p. 1.

11. In a legal sense, of course, the point is moot since Hitler declared war on the United States after the Japanese attack on Pearl Harbor. His declaration, however, came only after months of U.S. actions on behalf of Britain, Russia, and other allied powers that had made the United States a de facto belligerent.

12. Bruce M. Russett, *No Clear and Present Danger: A Skeptical View of the United States Entry into World War II* (New York: Harper and Row, 1972).

13. International Institute for Strategic Studies, *The Military Balance, 1993–1994* (London: Brassey's, 1993), p. 98.

14. Therese Raphael, "Yeltsin's Military Might," *Wall Street Journal*, September 23, 1993, p. A16. See also Ann Ignatius, "Russia Now Fields a Potemkin Military," *Wall Street Journal*, July 2, 1993, p. A4.

15. "Russia Mothballing 3 Carriers," *Washington Post*, February 15, 1994, p. A13.

16. Al Gore, "Forging a Partnership for Peace and Prosperity" (address before a conference sponsored by the Institute of World Affairs at the University of Wisconsin, Milwaukee, January 6, 1994), U.S. Department of State *Dispatch*, January 10, 1994, p. 15.

17. For a discussion of the distinction between vital and conditional (important but nevertheless secondary) security interests, see Ted Galen Carpenter, *A Search for Enemies: America's Alliances after the Cold War* (Washington: Cato Institute, 1992), pp. 170–79.

18. Quoted in Susanne Hoell, "Eastern Europe's Security Called Vital U.S. Interest," *Washington Times*, July 8, 1994, p. A16.

19. Richard Lugar, "NATO: Out of Area or Out of Business: A Call for U.S. Leadership to Revive and Redefine the Alliance" (remarks delivered to the Open Forum of the U.S. Department of State, August 2, 1993), pp. 5, 6.

20. Ibid., p. 4.

21. Ibid., p. 5.

22. Schwarz, p. 3.

## Chapter 3

1. See, for example, Christoph Bertram, "The New West, the New East," *Washington Post*, July 10, 1994, p. C7.

2. Ronald D. Asmus, Richard Kugler, and F. Stephen Larrabee, "Building a New NATO," *Foreign Affairs* 72 (September–October 1993): 37.

3. Stephen S. Rosenfeld, "No Threat to Moscow," *Washington Post*, July 8, 1994, p. A23.

4. Max M. Kampelman, foreword to *Searching for Moorings: East Central Europe in the International System*, ed. Jeffrey Laurenti (New York: United Nations Association of the United States, 1994), p. ix.

5. Steven Erlanger, "Russian Warning over Bigger NATO," *New York Times*, November 26, 1993, p. A11.

6. Ibid.

7. Craig R. Whitney, "NATO Bends to Russia to Allow It a Broader Partnership," *New York Times*, May 19, 1994.

8. William E. Schmidt, "Russia Tells NATO It Is Ready to Join Peace Partnership," *New York Times*, May 25, 1994, p. A1.

9. Steven Greenhouse, "Russia and NATO Agree to Closer Military Links," *New York Times*, June 23, 1994, p. A3.

10. Andrei Kozyrev, "Partnership for United, Peaceful, and Democratic Europe," *Frankfurter Rundschau*, January 8, 1994, translated text in *Foreign Broadcast Information Service Daily Report: Central Eurasia*, SOV-94-006, January 10, 1994, p. 7.

11. Ibid., p. 5.

12. Ibid.

13. Former vice president Alexander Rutskoi even denounced Russia's decision to join the Partnership for Peace. "It is a humiliating agreement," he fumed. "The 'Partnership for Peace' program is nothing but a means of making Russia submit to America's will." He added that Russia's role had become that of a "bootlicker." Interview with *L'Espresso* (Rome), reprinted in *Foreign Broadcast Information Service Daily Report: Central Eurasia,* SOV-94-136, July 15, 1994, p. 7.

14. For an overview of the attitudes of Russian officials, see Gerhard Wettig, "Moscow's Perception of NATO's Role," *Aussenpolitik* 2 (1994): 123–33.

15. Andrei Kozyrev, "The Lagging Partnership," *Foreign Affairs* 73 (May–June 1994): 65.

16. Ibid.

17. Asmus, Kugler, and Larrabee, p. 37.

18. Gary L. Geipel, "Expanding NATO: Good for Europe *and* Good for the U.S.," Hudson Briefing Paper no. 161, February 1994, p. 9.

19. Janusz Onyszkiewicz, "Why NATO Should Expand to the East," *Washington Post,* January 6, 1994, p. A27.

20. Edward L. Rowny, "NATO and the Difference between Eastern and Central Europe," *Washington Times,* March 15, 1994, p. A19.

21. Zbigniew Brzezinski, "The Premature Partnership," *Foreign Affairs* 73 (March–April 1994): 82.

22. Alexei Pushkov, "Building a New NATO at Russia's Expense," letter to the editor, *Foreign Affairs* 73 (January–February 1994): 173–74.

23. Andranik Migranyan, "Unequal Partnership," *New York Times,* June 23, 1994, p. A23.

24. Pushkov, p. 174.

25. Even a relative moderate like Georgi Arbatov complained that Yeltsin's foreign policy advisers were excessively compliant and "agreed to almost any proposal advanced by the West." Georgi Arbatov, "Eurasia Letter: A New Cold War?" *Foreign Policy* 95 (Summer 1994): 102.

26. For contrasting views about the significance of and motives underlying such moves, see Margaret Shapiro, "Russia Sees Itself as Region's Mediator," *Washington Post,* August 14, 1993, p. A17; Robert Cullen, "Russia Confronts Its 'Near Abroad,' " *Nation,* September 20, 1993, pp. 274–76; John P. Hannah, "The (Russian) Empire Strikes Back," *New York Times,* October 27, 1993, p. A23; Steven Erlanger, "Troops in Ex-Soviet Lands: Occupiers or Needed Allies?" *New York Times,* November 30, 1993, p. A1; and Thomas Goltz, "The Hidden Russian Hand," *Foreign Policy* 92 (Fall 1993): 92–116.

27. Warren Strobel, "Russia Abandons Arms Pact Revision," *Washington Times,* July 27, 1994, p. A14.

28. John Lloyd, "CIS Balks at Peacekeeping Bill," *Financial Times,* July 20, 1994, p. 2.

29. International Institute for Strategic Studies, *The Military Balance, 1993–1994* (London: Brassey's, 1993), pp. 93–98; Fred Hiatt, "Russia's Army: A Crumbling Giant," *Washington Post,* October 22, 1993, p. A1; Serge Schmemann, "A Once-Proud Force Finds Itself Impoverished and Demoralized," *New York Times,* November 28, 1993, p. A1;

Daniel Sneider, "Russian Budget Ignites Protest," *Defense News*, March 14–20, 1994, p. 3; and "Size of Army to Be Reduced," *Moscow Novaya*, reprinted in *Foreign Broadcast Information Service Daily Report: Central Eurasia*, SOV-94–128, July 5, 1994, p. 26.

30. Michael R. Gordon, "As Its World View Narrows, Russia Seeks a New Mission," *New York Times*, November 29, 1993, p. A1.

31. Even President Clinton seems surprisingly unwilling to placate Russia on that point. Addressing the Polish parliament, he assured his audience: "No democracy in this region should ever be consigned to a gray area, or a buffer zone. And no country should have the right to veto, compromise, or threaten democratic Poland's or any other democracy's integration into Western institutions, including those that ensure security." That statement would appear to reject any concept of even the most limited Russian sphere of influence in Eastern Europe. Bill Clinton, "Remarks by the President to the Sejm," White House, Office of the Press Secretary, July 7, 1994, p. 3.

32. An argument for a nuanced U.S. approach to the issue of a Russian sphere of influence can be found in Stephen Sestanovich, "Giving Russia Its Due," *National Interest* 36 (Summer 1994): 3–13.

33. Kampelman, p. ix.

34. For discussions, see William E. Schmidt, "Latvia's Worry: What to Do with All Its Russians," *New York Times*, March 1, 1994, p. A3; and Leyla Boulton, "Held Up on the Western Line," *Financial Times*, June 23, 1994, p. 16.

35. Steven Erlanger, "Latvia Amends Harsh Citizenship Law That Angered Russia," *New York Times*, July 24, 1994, p. A3. For a Russian criticism of Estonia's practices, see Boris Glebov, "Have They Forgotten about Human Rights?" *Rossiyskaya Gazeta*, July 19, 1994, reprinted in *Foreign Broadcast Information Service Daily Report: Central Eurasia*, SOV 94–139, July 20, 1994, pp. 9–10. Lithuania is the only Baltic republic to adopt a citizenship act that does not discriminate against Russian residents.

36. Bill Clinton, "Remarks at Freedom Monument," White House, Office of the Press Secretary, July 6, 1994, p. 4; and Thomas L. Friedman, "Clinton Makes Appeal to Latvia to Accept Its Russian Civilians," *New York Times*, July 7, 1994, p. A1.

37. Quoted in James Morrison, "Russian Reflex," *Washington Times*, July 22, 1994, p. A16.

38. Andrei Kozyrev, "Russia Plans Leading Role in World Arena," *Washington Times*, March 15, 1994, p. A14.

39. David Williams and R. Jeffrey Smith, "U.S. Intelligence Sees Economic Plight Leading to Breakup of Ukraine," *Washington Post*, January 25, 1994, p. A7. For general overviews of Ukraine's many problems and divisions, as well as its tense relations with Russia, see Paula J. Dobriansky, "Ukraine: A Question of Survival," *National Interest* 36 (Summer 1994): 65–72; Charles F. Furtado, Jr., "Nationalism and Foreign Policy in Ukraine," *Political Science Quarterly* 109 (Spring 1994): 81–104; and F. Stephen Larrabee, "Ukraine: Europe's Next Crisis?" *Arms Control Today* (July–August 1994): 14–19.

40. For a discussion of the various sources of tension between Russia and Ukraine, see William H. Kincaide and Natalie Melnyczuk, "Unneighborly Neighbors," *Foreign Policy* 94 (Spring 1994): 84–104.

41. Lee Hockstader, "Separatist Storm Brewing in Crimea," *Washington Post*, May 14, 1994, p. A16.

42. Mark Frankland, "Nuclear Neighbors Keep the Fences Up," *Washington Times*, June 11, 1993, p. A8.

43. For a discussion of Kiev's maneuvering and underlying motives, see Ted Galen Carpenter, "Staying Out of Potential Nuclear Crossfires," Cato Institute Policy Analysis no. 199, November 24, 1993, pp. 12–14.

44. Martin Sieff, "Ukraine's Stability in Danger," *Washington Times*, July 12, 1994, p. A1; and Anna Reid, "New Ukrainian Leader Reconsiders Nuclear Treaty," *Washington Post*, July 14, 1994, p. A16.

45. "Trilateral Statement by the Presidents of the United States, Russia and Ukraine," White House, Office of the Press Secretary, January 14, 1994, p. 2.

46. "U.S. Envoy Explains Security Guarantees," *Foreign Broadcast Information Service Daily Report: Central Eurasia*, SOV-94-014, January 21, 1994, p. 39.

47. Ann Devroy, "Pact Reached to Dismantle Ukraine's Nuclear Force," *Washington Post*, January 11, 1994, p. A1.

48. Ann Devroy and Daniel Williams, "Clinton Boosts A-Arms Pact in Ukraine," *Washington Post*, January 13, 1994, p. A1.

49. Ibid., p. A22.

50. Mona Charen, "Waving the Amulet of Disarmament," *Washington Times*, January 17, 1994, p. A17. Even the *Washington Post*, an advocate of both the Partnership for Peace and "incentives" to induce Kiev to relinquish its nuclear arsenal, expressed great uneasiness about the existence of secret provisions in the accord. "About That Ukraine Agreement," editorial, *Washington Post*, January 14, 1994, p. A22.

51. Charles L. Glaser, "Why NATO Is Still Best: Future Security Arrangements for Europe, " *International Security* 18 (Summer 1993): 10.

52. On the improbability of NATO's willingness to wage war to defend the nations of Central and Eastern Europe, see Stephen Blank, "New Challenges to European Security," *Strategic Review* (Summer 1994): 40–49.

53. William E. Odom, "Strategic Realignment in Europe: NATO's Obligation to the East," in *NATO: The Case for Enlargement* (London: Institute for European Defence and Strategic Studies, 1993), pp. 8–9.

54. Owen Harries, "The Collapse of 'The West,' " *Foreign Affairs* 72 (September–October 1993): 42.

55. Ibid., p. 43.

56. Kincaide and Melnyczuk, p. 102.

57. James A. Baker III, "A NATO Carrot to Solidify Reform," *Washington Times*, December 6, 1993, p. A18.

58. Leslie H. Gelb, "Can Clinton Deal with the World?" *Washington Post*, March 6, 1994, p. C1.

59. Zbigniew Brzezinski, "A Bigger—and Safer—Europe," *New York Times*, December 1, 1993, p. A23.

60. Brzezinski, "The Premature Partnership," pp. 81–82.

61. Jeane Kirkpatrick, "A Pillar for Democracies," *Washington Post*, December 6, 1993, p. A23.

62. "Include Russia, Too," editorial, *New York Times*, January 9, 1994, p. E20.

63. Kampelman, p. ix.

64. Coral Bell, "Why Russia Should Join NATO," *National Interest* 22 (Winter 1990–91): 37–47.

65. Henry Kissinger, "Not This Partnership," *Washington Post*, November 24, 1993, p. A17.

66. Alexander Haig, Jr., "Divots in the 'Partnership' Fairway," *Washington Times*, January 11, 1994, p. A14.

67. For discussions of the extent of the turmoil in Russia's near abroad, see Karen Dawisha and Bruce Parrott, *Russia and the New States of Eurasia: The Politics of Upheaval* (New York: Cambridge University Press, 1994); and Victor A. Kremenyuk, *Conflicts in and around Russia: Nation-Building in Difficult Times* (Westport, Conn.: Greenwood, 1994).

68. Kissinger, "Not This Partnership."

69. Migranyan.

## Chapter 4

1. Charles L. Glaser, "Why NATO Is Still Best: Future Security Arrangements for Europe," *International Security* 18 (Summer 1993): 7.

2. Josef Joffe, "America's European Pacifier," *Foreign Policy* 54 (Spring 1984): 64–82. See also Josef Joffe, "NATO and the Limits of Devolution," in *NATO at 40: Confronting a Changing World*, ed. Ted Galen Carpenter (Lexington, Mass.: Lexington Books, 1990), pp. 59–74.

3. John J. Mearsheimer, "Back to the Future: Instability in Europe after the Cold War," *International Security* 15 (Summer 1990): 5–56. Mearsheimer is pessimistic that even NATO can prove effective in that role in the long term, as the alliance and other Cold War institutions become less relevant and robust in the post–Cold War era.

4. Glaser, p. 21.

5. Bill Clinton, "Remarks of the President to the Sejm," White House, Office of the Press Secretary, July 7, 1994, p. 4.

6. Vaclav Havel, "New Democracies for Old Europe," *New York Times*, October 17, 1993, p. E17.

7. Ibid.

8. Ibid.

9. "Excerpts from President Clinton's News Conference," *Washington Post*, June 18, 1993, p. A10.

10. Earl C. Ravenal, *Designing Defense for a New World Order* (Washington: Cato Institute, 1991), p. 1.

11. William Pfaff, "NATO Has a Big Role to Play in Europe's Security," *Chicago Tribune*, May 2, 1993, p. D3. For a more detailed exposition of his thesis, see William Pfaff, "Invitation to War," *Foreign Affairs* 72 (Summer 1993): 97–109.

12. Al Gore, "Forging a Partnership for Peace and Prosperity" (address before a conference sponsored by the Institute of World Affairs at the University of Wisconsin, Milwaukee, January 6, 1994), U.S. Department of State *Dispatch*, January 10, 1994, p.15.

13. Pat Buchanan, "Leading Us Down a Primrose Path," *Washington Times*, January 14, 1994, p. A19.

14. Quoted in David B. Ottaway, "Ethnic Hungarians in Slovakia Are Demanding Self-Government," *Washington Post*, January 10, 1994, p. A12.

15. The new Socialist party government of Prime Minister Gyula Horn has pledged to pursue a less nationalist foreign policy and to seek a "historic rapprochement" with its neighbors. Whether that intention is sincere and, even if it is, whether Horn can make good on the promise, remains to be seen. David B. Ottaway, "Hungary May Seek Detente with Neighbors," *Washington Post*, May 29, 1994, p. A51.

16. David B. Ottaway, "A Battle for Identity Divides Transylvania," *Washington Post*, July 5, 1994, p. A12.

17. Jonathan Sunley, "Hungary Spurns the Neo-communist Urge—So Far," *Wall Street Journal*, October 11, 1993, p. A12, European edition.

18. Quoted in David B. Ottaway, "Hungary Bans NATO Planes That Would Assist Airstrikes in Bosnia," *Washington Post*, February 15, 1994, p. A14.

19. Quoted in Judy Dempsey, "Kozyrev Warns E. Europe on NATO," *Financial Times*, February 21, 1994, p. 2.

20. For discussions, see Ted Galen Carpenter, "U.S. Troops in Macedonia: Back Door to War?" Cato Institute Foreign Policy Briefing no. 30, March 17, 1994, pp. 3–5; and Edward Mortimer, "Southern Discomfort," *Financial Times*, March 3, 1994, p. 13.

21. James Rupert, "Athens Adds to Embargo of Macedonia," *Washington Post*, February 19, 1994, p. A22; Lionel Barber, "Greece Warned on 'Illegal' Embargo," *Financial Times*, February 22, 1994, p. 1; George Soros, "The Other Balkan Mess," *New York Times*, March 17, 1994, p. A23; and Henry Kamm, "Greek Prime Minister Insists Macedonia Endangers His Country," *New York Times*, April 7, 1994, p. A11.

22. Roger R. Cohen, "Macedonia Census Just Inflames the Disputes," *New York Times*, July 17, 1994, p. A8.

23. Andrew Borowiec, "Macedonia Says It Has Thwarted a Military Plot from Albania," *Washington Times*, November 11, 1993, p. A15.

24. Henry Kamm, "Macedonia Sees Its Albanians as Its 'Biggest Problem,' " *New York Times*, May 5, 1994, p. A15.

25. George Jahn, "Albania Inches Ahead with U.S. as Suitor," *Washington Times*, May 10, 1994, p. 14.

26. Quoted in "Washington Whispers," *U.S. News & World Report*, April 25, 1994, p. 27.

27. Quoted in Jahn.

28. Monteagle Stearns, *Entangled Allies: U.S. Policy toward Greece, Turkey, and Cyprus* (New York: Council on Foreign Relations, 1992).

29. Ibid. The massive inflow of armaments into both countries from the United States, Germany, and other NATO members during the past three years makes an already dangerous situation even worse. Bruce Clark, "NATO Arms Pour into Greece and Turkey," *Financial Times*, June 7, 1994, p. 2.

30. See, for example, Kevin F. Donovan, "The American Response to European Nationalism," in *NATO's Eastern Dilemma*, ed. David G. Haglund, S. Neil MacFarlane, and Joel J. Sokolsky (Boulder, Colo.: Westview, 1994), pp. 95–96.

31. Caryle Murphy, "Turkey Accused of Rights Abuses in Effort to Crush Kurdish Rebels," *Washington Post*, May 12, 1994, p. A20.

32. "Tragedy in Turkey," editorial, *New York Times*, April 4, 1994, p. A14.

33. John M. Goshko, "Bush Threatens 'Military Force' if Serbs Attack Ethnic Albanians," *Washington Post*, December 29, 1992, p. A10; and "Clinton Warns Serbian Leaders on Military Action in Kosovo," *Washington Post*, March 2, 1993, p. A14.

34. Barton Gellman and Ann Devroy, "U.S. Weighs Troops for Macedonia," *Washington Post*, May 12, 1993, p. A1.

35. Margaret Thatcher, "Stop the Serbs. Now. For Good," *New York Times*, May 4, 1994, p. A23.

36. Robert W. Tucker and David C. Hendrickson, "America and Bosnia," *National Interest* 33 (Fall 1993): 17.

37. Nikolaos Stavrou, "The Balkan Quagmire and the West's Response," *Mediterranean Quarterly* 4 (Winter 1993): 36.

38. David B. Ottaway, "Yugoslavia Stands by Prior Demand That Bosnian Serbs Give Up Land," *Washington Post*, April 29, 1994, p. A45. On other aspects of the friction, see Laura Silber, "Bosnian Serbs Test Patience of Milosevic," *Financial Times*, May 4, 1994, p. 2; and Aleksa Djilas, "Serbs vs. Serbs in Bosnia, Belgrade," *New York Times*, July 21, 1994, p. A23.

39. William Safire, "General Shilly-Shali," *New York Times*, April 21, 1994, p. A19.

**Chapter 5**

1. John F. Harris, "Perry Sees Wider Role in Bosnia," *Washington Post*, July 18, 1994, p. A1; and Bruce Clark, "U.S. and NATO Likely to Expand Role in Bosnia," *Financial Times*, July 19, 1994, p. 2.

2. Anthony Lake, "Bosnia, America's Interests and America's Role" (remarks at Johns Hopkins University, Baltimore, Maryland, April 7, 1994), p. 2.

3. Barton Gellman, "For NATO, a First That Took 45 Years," *Washington Post*, March 1, 1994, p. A13.

4. Michael R. Gordon, "U.S. Rules Out Military Force against Serbs," *New York Times*, April 4, 1994, p. A1.

5. Craig R. Whitney, "NATO Sees a Role with Peacekeepers from Eastern Europe," *New York Times*, June 5, 1992, p. A1; and William Drozdiak, "NATO Widens Mandate on Forces," *Washington Post*, June 5, 1992, p. A41.

6. NATO increasingly sees itself acting on behalf of the United Nations in Bosnia-style conflicts and is reconfiguring many of its forces for peacekeeping operations rather than traditional territorial defense missions. Brooks Tigner, "NATO Eyes Peacekeeping Tools," *Defense News*, July 11–17, 1994, p. 4. For a more general discussion, see J. M. Boorda, "Loyal Partner: NATO's Forces in Support of the United Nations," *NATO's Sixteen Nations* 1 (1994): 8–12.

7. Michael R. Gordon, "NATO and the U.N. in Dispute over Strikes against the Serbs," *New York Times*, April 24, 1994, p. A12.

8. Lake, p. 5.

9. Roger Cohen, "Washington Might Recognize a Bosnian Serb State," *New York Times*, March 13, 1994, p. A10. Emphasis added.

10. Ibid.

11. Lake, p. 6.

12. Ibid., pp. 6–7. A few months later, Secretary of Defense Perry reiterated that the United States was ready to send "a significant number" of troops to Bosnia to implement a peace accord. Quoted in Clark, p. 2.

13. John Pomfret, "UN Tanks Kill 9 Serbs in Bosnia," *Washington Post*, May 2, 1994, p. A1.

14. John McCain, "In Bosnia, Another Mistake," *Wall Street Journal*, April 15, 1994, p. A8.

15. "A Craving for Peace—and for More War," *U.S. News & World Report*, May 9, 1994, p. 13.

16. International Institute for Strategic Studies, *The Military Balance, 1992–1993* (London: Brassey's, 1992), pp. 51, 87.

17. Daryl G. Press, "What's Happening in Bosnia Is Terrible, but It's Not Genocide," *Christian Science Monitor*, September 26, 1993, p. E4.

18. Furthermore, ethnically motivated killings have not been confined to Serb forces. Croat and Bosnian Muslims have also committed such atrocities, although that fact has generally been lost in the barrage of stories about Serb crimes. See, for example, John F. Burns, "In Muslim Town, Serbs Pay the Price," *New York Times*, March 9, 1993, p. A6.

19. Andrew Bell-Fialkoff, "A Brief History of Ethnic Cleansing," *Foreign Affairs* 72 (Summer 1993): 110–21.

20. Larry Collins and Dominique Lapierre, *Freedom at Midnight* (New York: Simon and Schuster, 1975), pp. 329–98. A similar although less bloody expulsion of populations occurred following the UN-mandated partition of Palestine into Arab and Jewish states. Some 700,000 Palestinians ultimately fled their homes—many in response to terrorist campaigns conducted by Israeli paramilitary organizations.

21. Alfred M. de Zayas, *Nemesis at Potsdam: The Anglo-Americans and the Expulsion of the Germans* (London: Routledge & Kegan Paul, 1977).

22. "More Than Bosnia," editorial, *Wall Street Journal*, April 28, 1994, p. A12.

23. Dov Zakheim, "A Syndrome That Was Only in Remission?" *Washington Times*, May 3, 1994, p. A17.

24. Margaret Thatcher, "Stop the Serbs. Now. For Good," *New York Times*, May 4, 1994, p. A23.

25. "More Than Bosnia."

26. Christopher Layne and Benjamin Schwarz, "American Hegemony—Without an Enemy," *Foreign Policy* 92 (Fall 1993): 16.

27. Ibid.

28. Fareed Zakaria, "Bosnia Explodes Three Myths," *New York Times*, September 26, 1993, p. E15.

29. Paul Greenberg, "Where They Never Learn," *Washington Times*, June 3, 1992, p. G3. That point also seems to impress Anthony Lewis, who observes that "this terrible century began its downward slide 78 years ago in Sarajevo." Anthony Lewis, "The New World Order," *New York Times*, May 17, 1992, p. E17.

30. George Melloan, "Gorazde's Lessons about Foreign Policy Drift," *Wall Street Journal*, April 25, 1994, p. A15.

31. Benjamin Schwarz, "Leave the Little Wars Alone," *Los Angeles Times*, June 8, 1992, p. B-5.

32. That danger has increased with the introduction of Turkish military units as part of the UN-directed peacekeeping operation. Having Turkey, which consistently backs the Muslim faction in Bosnia, and Russia, with its pronounced pro-Serb bias, both involved militarily creates what journalist John Pomfret aptly describes as "a potentially explosive brew." John Pomfret, "U.N. Tells Troops from Turkey to Keep Low Profile in Bosnia," *Washington Post*, July 16, 1994, p. A15.

33. Patrick Buchanan, "Power Rivalries Revisited," *Washington Times*, February 23, 1994, p. A16.

34. Accounts of the widespread anger in the Russian parliament, including that of reform deputies, at NATO's increasingly coercive policy in Bosnia can be found in Steven Erlanger, "Anti-Western Winds Gain Force in Russia," *New York Times*, April 17, 1994, p. E4.

35. Lee Hockstader, "Yeltsin Voices Anger That He Was Not Consulted on Raids," *Washington Post*, April 12, 1994, p. A14.

36. Steven Greenhouse, "U.S., Britain and Russia Form Group to Press Bosnia Accord," *New York Times*, April 26, 1994, p. A6.

37. The Russians have also exploited opportunities to flex their diplomatic muscles on peripheral issues concerning the Bosnian conflict. For example, Russian officials vetoed the appointment of a jurist from any NATO country as the prosecutor who will bring cases before the UN tribunal that is tasked with prosecuting Balkan war crimes. Moscow claimed that appointees from NATO members would be biased against the Serbs. Paul Lewis, "South African Is to Prosecute Balkan War Crimes," *New York Times*, July 9, 1994, p. A2.

38. Celestine Bohlen, "Russian General Opposes Air Strikes on Serbs," *New York Times*, April 26, 1994, p. A7. See also, Lee Hockstader, "Divided Russian Officials Publicly Squabble over Pro-Serb Policy," *Washington Post*, April 24, 1994, p. A26.

39. Alan Riding, "Western Nations to Add Sanctions after Serbs Balk," *New York Times*, July 31, 1994, p. A1; and Daniel Williams, "Meeting on Bosnia Yields 'Lowest Common Denominator' of Agreement," *Washington Post*, August 1, 1994, p. A17.

40. For a discussion of how NATO is ill-equipped, temperamentally and institutionally, for such missions, see Charles William Maynes, "NATO's Tough Choice in Bosnia," *New York Times*, July 27, 1994, p. A21.

## Chapter 6

1. Catherine McArdle Kelleher, "Cooperative Security in Europe," in *Global Engagement: Cooperation and Security in the 21st Century*, ed. Janne E. Nolan (Washington: Brookings Institution, 1994), p. 316.

2. Robert D. Hormats, "Redefining the Atlantic Link," *Foreign Affairs* 68 (Fall 1989): 86.

3. David M. Abshire, "Don't Muster Out NATO Yet: Its Job Is Far from Done," *Wall Street Journal*, December 1, 1989, p. A14.

4. Daniel Graham, "NATO and the Spread of Ballistic Missiles," *Washington Times*, November 16, 1991, p. D8.

5. Richard Nixon, *Seize the Moment: America's Challenge in a One-Superpower World* (New York: Simon and Schuster, 1992), pp. 142–43.

6. Warren Christopher, "Strengthening the Atlantic Alliance through a Partnership for Peace" (remarks at the North Atlantic Council Ministerial Meeting, NATO Headquarters, Brussels, December 2, 1993), U.S. Department of State *Dispatch*, December 13, 1993, p. 857.

7. Quoted in Craig Covault, "NATO Eyes Protection for Military Transports," *Aviation Week and Space Technology*, March 14, 1994, p. 52.

8. Leslie H. Gelb, "Can Clinton Deal with the World?" *Washington Post*, March 6, 1994, p. C1.

9. Quoted in Douglas Jehl, "CIA Nominee Wary of Budget Cuts," *New York Times*, February 3, 1993, p. A18.

10. David S. Broder, "Countering Critics, Defending Decisions," *Washington Post*, May 12, 1994, p. A11.

11. For a discussion of the various problems and dangers associated with economic espionage, see Stanley Kober, "The CIA as Economic Spy: The Misuse of U.S. Intelligence after the Cold War," Cato Institute Policy Analysis no. 185, December 8, 1992.

12. Robert Dreyfuss, "Company Spies," *Mother Jones*, May–June 1994, pp. 16–17.

13. Quoted in David Isenberg, "The Pentagon's Fraudulent Bottom-Up Review," Cato Institute Policy Analysis no. 206, April 21, 1994.

14. Quoted in Jim Wolffe, "Powell Sees Opportunity for U.S. to Reduce Military Strength," *Defense News*, April 8, 1991, p. 12. Contrary to the title of the article, Powell did not really regard reduced budgets as an "opportunity." Rather, he saw them as an inevitable result of the changing international security environment and domestic political pressures.

15. James M. Buchanan and Gordon Tullock, *The Calculus of Consent* (Ann Arbor: University of Michigan Press, 1965), p. 286.

16. For a discussion of issue networks and their roles, see Hugh Heclo, "Issue Networks and the Executive Establishment," in *Public Administration: Concepts and Cases*, 4th ed., ed. Richard J. Stillman II (Boston: Houghton Mifflin, 1988), pp. 408–17.

17. Lawrence J. Korb, "Shock Therapy for the Pentagon," *New York Times*, February 15, 1994, p. A21.

18. Richard G. Lugar, "NATO's Near Abroad: New Membership, New Missions" (speech to the Atlantic Council, December 9, 1993), p. 18.

19. The title of Brzezinski's latest book is rather revealing in that regard. Zbigniew Brzezinski, *Out of Control: Global Turmoil on the Eve of the 21st Century* (New York: Collier Books, 1993). The tone throughout suggests that the disorder associated with global kaleidoscopic change automatically poses a serious threat to American interests, values, and well-being.

20. For a discussion of the minimal effect of such wars on U.S. security, see Barbara Conry, "The Futility of U.S. Intervention in Regional Conflicts," Cato Institute Policy Analysis no. 209, May 19, 1994.

21. For a more detailed exposition of Lake's Wilsonian views, see Anthony Lake, "From Containment to Enlargement" (remarks at John's Hopkins University School of Advanced International Studies, Washington, September 21, 1993). For similar views by another high-ranking administration official, see Madeleine K. Albright, "Remarks to the National War College, National Defense University, Washington," September 23, 1993.

## Chapter 7

1. Jonathan G. Clarke, "Replacing NATO," *Foreign Policy* 93 (Winter 1993–94): 23.

2. The decision by the other CSCE members to "suspend" Yugoslavia in June 1992 appears to have been a ploy to get around the unanimity requirement and its virtual guarantee of deadlock. It is not yet clear what the implications of that precedent will be for the effectiveness of the organization.

3. Ronald Steel, "NATO's Afterlife," *New Republic*, December 2, 1991, p. 18.

4. *Platform on European Security Interests* (The Hague: Western European Union, October 27, 1987).

5. Geoffrey Howe, "The Atlantic Alliance and the Security of Europe," *NATO Review* 35 (April 1987): 8.

6. For discussions of those developments and the role they played in embryonic efforts to create a distinctly European security identity, see David Garnham, "U.S. Disengagement and European Defense Cooperation," in *NATO at 40: Confronting a Changing World*, ed. Ted Galen Carpenter (Lexington, Mass.: Lexington Books, 1990), pp. 75–91; and David Garnham, *The Politics of European Defense Cooperation: Germany, France, Britain and America* (Cambridge, Mass.: Ballinger, 1988).

7. Jonathan G. Clarke, "The Eurocorps: A Fresh Start in Europe," Cato Institute Foreign Policy Briefing no. 21, December 28, 1992.

8. Dudley Smith, introduction to "The WEU and the European Common Foreign and Security Policy," Institut Royal Supérieur de Défense, Brussels, October 5, 1993, p. 2. For a discussion of EU developments in the political and security realms, see Desmond Dinan, *Ever Closer Union* (Boulder, Colo.: Westview, 1994).

9. "State of the World in the 1990s," U.S. Institute of Peace *Journal*, February 1991, p. 3.

10. "Eurocorps Will Strengthen European Pillar of NATO," press release, (German) Federal Press Office, July 22, 1992, p. 15.

11. Clarke, "Eurocorps," p. 4.

12. Quoted in Frederick Kempe, "U.S., Bonn Clash over Pact with France," *Wall Street Journal*, May 27, 1992, p. A9.

13. Quoted in William Drozdiak and Ann Devroy, "Bush Challenges Europeans to Define U.S. Role," *Washington Post*, November 8, 1991, p. A1.

14. Warren Christopher, "Strengthening the Atlantic Alliance through a Partnership for Peace" (remarks at the North Atlantic Council Ministerial Meeting, NATO Head-quarters, Brussels, December 2, 1993), U.S. Department of State *Dispatch*, December 13, 1993, p. 858.

15. NATO Declaration, January 11, 1994, U.S. Department of State *Dispatch* 5, supplement no. 1 (January 1994): 7.

16. Ibid., p. 8.

17. Ibid.

18. Ibid.

19. Don Cook, *Forging the Alliance* (New York: Arbor House, 1989), pp. 114–16.

20. Josef Joffe, "Collective Security and the Future of Europe," *Survival* 34 (Spring 1992): 47.

21. Jeffrey Simon, "Does Eastern Europe Belong in NATO?" *Orbis* 37 (Winter 1993): 24.

22. Some cautious expansion already appears to be under way. The WEU has offered associate status to the Baltic republics and other Central and East European states, and France and Germany have invited Poland to send observers to the Eurocorps as a prelude to Poland's eventual full membership in that organization. "Justice Minister: Joining WEU Means Closer Participation," *Foreign Broadcast Information Service Daily Report: Central Eurasia*, SOV-095, May 17, 1994, p. 72; and Giovanni de Briganti, "Eurocorps May Include Polish Troops," *Defense News*, March 7–13, 1994, p. 1.

23. Quoted in Daniel Williams, "Western Envoys Wary of Russia's Entry into NATO," *Washington Post*, June 25, 1994, p. A18. For the views of the Russian foreign ministry, see "Dissatisfaction Voiced with WEU Actions on Partnership," *Foreign Broadcast Information Service Daily Report: Central Eurasia*, SOV-093, May 13, 1994, p. 13.

24. "Justice Minister," p. 72.

25. Steven L. Canby, "We Are Missing a Golden Opportunity to Secure Western Europe," *Washington Times*, February 26, 1994, p. D3.

26. Brooks Tigner, "Possibilities Grow Dim for E. Europe Alliance," *Defense News*, March 7–13, 1994, p. 3.

27. Ronald D. Asmus, Richard L. Kugler, and F. Stephen Larrabee, "Building a New NATO," *Foreign Affairs* 72 (September–October 1993): 35.

28. Richard Lugar, "NATO: Out of Area or Out of Business: A Call for U.S. Leadership to Revive and Redefine the Alliance" (remarks delivered to the Open Forum of the U.S. Department of State, August 2, 1993), p. 7.

29. Steel, p. 18.

30. Quoted in Andrew Borowiec, "Clinton and Europe: The Morning After," *Washington Times*, January 19, 1994, p. A16.

31. Jeane Kirkpatrick, "An Active Europe, a Passive United States," *Washington Post*, November 25, 1991, p. A21.

32. For discussions, see David P. Calleo, *Beyond American Hegemony: The Future of the Atlantic Alliance* (New York: Basic Books, 1987), passim; and Alan Tonelson, "The Economics of NATO," in *NATO at 40*, pp. 93–107.

33. "Thatcher Ruthless with U.S. in 1985," *Washington Times*, March 16, 1994, p. A12.

34. Quoted in Philip Finnegan, "Perry Raps Budget Focus," *Defense News*, July 25–31, 1994, p. 1.

35. Quoted in ibid.

36. For a discussion of Washington's growing difficulties in exercising hegemony in the alliance, see Joseph Lepgold, *The Declining Hegemon: The United States and European Defense, 1960–1990* (New York: Praeger, 1990).

37. Owen Harries, "The Collapse of 'The West,' " *Foreign Affairs* 72 (September–October, 1993): 41–42.

38. Ibid., p. 48.

39. Quoted in Stanley R. Sloan, "The NATO Summit: Transatlantic Relations at a Crossroads," Congressional Research Service Report for Congress, October 28, 1993, p. 5.

40. Clarke, "Eurocorps," p. 8.

41. Ibid.

# Index

# About the Author

Ted Galen Carpenter is director of foreign policy studies at the Cato Institute. He is the author of *A Search for Enemies: America's Alliances after the Cold War* and the editor of four books on defense and foreign policy issues, including *NATO at 40: Confronting a Changing World.*

He has contributed chapters to 14 books on international relations, and his articles have appeared in numerous policy journals including *Foreign Affairs, Foreign Policy, World Policy Journal, International History Review, National Interest,* and *Politique Internationale.*

Carpenter received his B.A. and M.A. in U.S. history from the University of Wisconsin at Milwaukee and his Ph.D. in U.S. diplomatic history from the University of Texas.

# Cato Institute

Founded in 1977, the Cato Institute is a public policy research foundation dedicated to broadening the parameters of policy debate to allow consideration of more options that are consistent with the traditional American principles of limited government, individual liberty, and peace. To that end, the Institute strives to achieve greater involvement of the intelligent, concerned lay public in questions of policy and the proper role of government.

The Institute is named for *Cato's Letters*, libertarian pamphlets that were widely read in the American Colonies in the early 18th century and played a major role in laying the philosophical foundation for the American Revolution.

Despite the achievement of the nation's Founders, today virtually no aspect of life is free from government encroachment. A pervasive intolerance for individual rights is shown by government's arbitrary intrusions into private economic transactions and its disregard for civil liberties.

To counter that trend, the Cato Institute undertakes an extensive publications program that addresses the complete spectrum of policy issues. Books, monographs, and shorter studies are commissioned to examine the federal budget, Social Security, regulation, military spending, international trade, and myriad other issues. Major policy conferences are held throughout the year, from which papers are published thrice yearly in the *Cato Journal*. The Institute also publishes the quarterly magazine *Regulation*.

In order to maintain its independence, the Cato Institute accepts no government funding. Contributions are received from foundations, corporations, and individuals, and other revenue is generated from the sale of publications. The Institute is a nonprofit, tax-exempt, educational foundation under Section 501(c)3 of the Internal Revenue Code.

CATO INSTITUTE
1000 Massachusetts Ave., N.W.
Washington, D.C. 20001